Technology and Nationalism in India

Technology and Nationalism in India

Cultural Negotiations From Colonialism to Cyberspace

Rohit Chopra

CAMBRIA
PRESS
Amherst, New York

Requests for permission should be directed to:
permissions@cambriapress.com, or mailed to:
Cambria Press
20 Northpointe Parkway, Suite 188
Amherst, NY 14228

Library of Congress Cataloging-in-Publication Data

Chopra, Rohit.
Technology and nationalism in India : cultural negotiations from colonialism to cyberspace / Rohit Chopra.
p. cm.
Includes bibliographical references and index.
ISBN 978-1-60497-567-3 (alk. paper)
1. Nationalism and technology—India—History. 2. Hindutva. 3. Internet—Social aspects—India. I. Title.

DS428.2.C46 2008
320.540954—dc22

2008038073

Table of Contents

List of Figures

List of Tables

Acknowledgments

I.

I have the pleasurable and impossible task of adequately thanking all those who, in different and distinct ways, have made this book possible.

This book has been developed from my dissertation at the Graduate Institute of the Liberal Arts (ILA), a unique space for interdisciplinary inquiry and scholarship at Emory University in Atlanta, Georgia. I have been fortunate to have benefited from the expertise and learning of numerous people in the ILA and other departments at Emory as a graduate student and, later, as a visiting assistant professor in 2006–2007. I wish to express my profound gratitude to Allen Tullos and Cynthia Patton for their intellectual guidance, mentorship, and encouragement over the years. I am enormously grateful to Irene Browne, Deepika Bahri, and Laurie Patton for the privilege of an apprenticeship with each of them. I have gained greatly from discussions with Angelika Bammer, who has also been a strong source of guidance and encouragement. They have all helped this book take shape and evolve from an idea to its present form. I would like to thank Edna Bay, Alan Cienki, John Johnston, Dalia Judovitz, Ivan Karp, and Frank Lechner, conversations with whom in the course of my time at Emory have enriched the book. I also wish to express my thanks to Kevin Corrigan, Cris Levenduski, Anita Mann, and Walt Reed at Emory for their support.

I owe a special debt to Abdullahi An-Na'im. It has been my privilege to have worked with him on a number of projects at the Center for the Study of Law and Religion at Emory Law School, an immensely rewarding experience that has strongly influenced the course of my research. I am no less indebted to him for his guidance as, indeed, for the inspiration of his example.

Daniel Miller at University College, London, kindly took time out to meet me and speak with me about my research during the early stages of the project. My former colleagues at the History and Society and Arts and Humanities divisions at Babson College, Massachusetts, where I taught in 2006–2007 as an assistant professor of media studies, were generous in their support for the project. I am honored to be a part of the Department of Communication at Santa Clara University, where this book has come to fruition. I am extremely grateful to my colleagues in the department and to the university for providing an atmosphere of intellectual vitality, collegiality, and encouragement that enables scholarship to flourish. I would be remiss if I did not thank Toni Tan, Kristin Roth, Matthew Sullivan, and others at Cambria Press. It has been a pleasure working with them.

Across three continents, there are those who have shared the gift of friendship over the years: Antonio de Velasco, who has also been an important intellectual interlocutor at every stage of the book; Debadeep Bhattacharyya and Dipanwita Bhattacharyya; Susan Chen; Marsha Ford; Arjun Lahiri and Komal Lahiri; Rakesh Ranjan; Subhrasheel Roychowdhury and Laura Bear; and Mary Alma Durrett.

And then there are those debts that are beyond words. This book could not have been written without the love, warmth, and good cheer of Jagat Pal Chopra and Urvashi Chopra, Rahul Chopra and Maya Dodd, Govind

Shahani and Roshan Shahani, and Nishant Shahani. It is my regret that Mataji and Pitaji, Nana and Nani, and Amma did not live to see this book.

Most of all, for everything, I thank Gitanjali Shahani.

II.

Parts of this book were first published in the following sources. I am grateful to the publishers for kindly granting me permission to include these materials here.

- "Neoliberalism as Doxa: Bourdieu's Theory of the State and the Contemporary Indian Discourse on Globalization and Liberalization," *Cultural Studies* 17, nos. 3–4 (2003): 419–444, http://informaworld.com.
- "Global Primordialities: Virtual Identity Politics in Online Hindutva and Online Dalit Discourse." The final definitive version of this paper has been published in *New Media and Society* 8, no. 2 (2006): 187–206 by Sage Publications Ltd., all rights reserved. © 2006, Sage Publications Ltd., http://nms.sagepub.com.
- "The Cyber-Presence of Babri Masjid: History, Politics, and Difference in Online Indian Islam," *Economic and Political Weekly* 43, no. 3 (January 19–25, 2008): 47–56, http://www.epw.org.in.
- "The Virtual State of the Nation: Online Hindu Nationalism in Global Capitalist Modernity," in *South Asian Technospaces*, ed. Radhika Gajjala and Venkataramana Gajjala (New York: Peter Lang, 2008), 153–178.

Technology and Nationalism in India

Chapter 1

Introduction: Technology and Nationalism in India From Colonialism to Cyberspace

It was science alone that could solve these problems of hunger and poverty, of insanitation and illiteracy, of superstition and deadening custom and tradition, of vast resources running to waste, of a rich country inhabited by starving people.

—Jawaharlal Nehru[1]

Arithmetic, the number, have always had a decisive role in the State apparatus: this is so even as early as the imperial bureaucracy... It is even more true of modern forms of the State, which in developing utilized all the calculation techniques that were springing up at the border between mathematical science and social technology (there is a whole social calculus at the basis of political economy, demography, the organization of work, etc.). This arithmetic

> *element of the State found its specific power in the treatment of all kinds of matter: primary matter (raw materials), the secondary matter of worked objects, or the ultimate matter constituted by the human population.*
>
> —Gilles Deleuze and Félix Guattari[2]

This book examines the use of the Internet by global Indian communities for the promotion of Hindu nationalist ideologies, a phenomenon that I term "technocultural Hindu nationalism," with reference to the relationship between technology and nationalism in India from the period of British colonial rule in the mid-eighteenth century to the present era of an economically and technologically interconnected world. Since the introduction of Western science and technology under colonial rule in India in the eighteenth century, science and technology have been used as instruments of mapping, controlling, and transforming Indian society. Equally, scientific and technological expertise have been authorized as attributes of modern Indian selfhood—as markers, at once, of a universality and an essential Indianness. Since the colonial era, in the imagination of both state and society, the possessors of technological skills have been vested with the authority to speak for the nation. State and society have granted the holders of technological skills and qualifications economic opportunities, educational advantages, and social status. Even as the relationship between technology and nationalism has been marked by continuities and discontinuities over the passage of time, it has endured as fundamentally constitutive of Indian modernity.

As Gyan Prakash has compellingly demonstrated, in the nineteenth century, an upper-caste educated Hindu intelligentsia argued that the defining characteristic of ancient Indian civilization was its scientific and technological character—one that could be mobilized to effect an Indian renaissance and liberate India from the effects of Islamic invasions and British colonial domination.[3] In elite Indian discourse, scientific expertise was valorized as a traditional Hindu attribute and the means to a unique Indian modernity. This conception of scientificity was inextricably woven into emergent narratives of anticolonial nationalism. With independence in 1947, in keeping with the model of the Indian nation-state envisioned by Jawaharlal Nehru, science and technology in tandem with official philosophies of socialism and secularism were entrusted with the task of nation building. It was hoped that ambitious state projects of technological development would lead to economic prosperity and social justice for all segments of Indian society. The idea of the scientific temperament as a highly desirable individual and collective national characteristic was also strongly reinforced through state policies. Indian socioeconomic elites selectively internalized this statist discourse, for example, by delinking the imperative of social justice from the acquisition of scientific and technological skills. In 1991 the Indian state, through policy changes in favor of globalization and free market economic reforms, adopted economic neoliberalism as the official framework for governance and national development. The 1980s and 1990s also witnessed the resurgence of the ideology of Hindu nationalism or Hindutva—the idea that India is essentially a Hindu nation and that minorities, particularly Muslims and Christians, are outsiders—in Indian society. The domain of technology, especially the Internet, is a key area where these respective perspectives on the Indian nation converge. On Hindu nationalist Web sites addressed at a global Hindu audience, Hinduism is defined as ideally suited to the demands of a global capitalist modernity exemplified by free market ideologies and technologies such as the Internet. On these Web sites, Hindu religion and culture are also emphasized as the authentic foundations of Indian national identity. Technology, particularly the Internet, enables and symbolizes the conflation of the categories of Indian and Hindu into a new articulation of national identity and nationalism.

This book describes the historical negotiation between varied, often oppositional, discourses of technology and nationalism from the colonial era until the Internet age, with the objective of illuminating the complex dynamics of Hindu nationalist identity politics in cyberspace. This inquiry involves addressing four broad, related questions:

1. What are the defining characteristics of the relationship between technology and nationalism in different historical phases (the era

of anticolonial nationalism, the period of Nehruvian socialism, and the present context of globalization and Hindu nationalism)?
2. How do these multiple legacies inform technocultural Hindu nationalism?
3. What is the consequence of the intersection of Hindu nationalist discourse and the Internet when compared to other Indian nationalisms as these manifest themselves online?
4. How do the modalities of Internet communication contribute to the forms taken by Hindu nationalism in cyberspace?

A description of technocultural Hindu nationalism, with reference to its immediate and longer histories, will illuminate the salience of these concerns.

To the observer of Indian affairs, the twenty-first century presents two strikingly contradictory sets of images about the Indian state and society. India simultaneously appears as a rising economic star in the international firmament and a politically unstable and fragile society, at once the vital technology hub of the global future and a land devastated by primordial religious and nationalist conflicts, a nation both poised to play an active role in world matters in the new millennium and condemned to infinitely repeating the traumas of her past. On the one hand, the Indian state and citizens exemplify scientific and technological prowess coupled with economic enterprise—qualities that optimally place the country and people to reap the rewards of global capitalist modernity. On the other hand, the Indian nation and inhabitants seem to stand for chauvinistic nationalism and religious discord—atavistic burdens that bode ill for the economic, political, and social future of the country.

For the last several years, the diasporic Indian media, at least its English-language component, has been churning out stories on the rise of the nation as a new global economic and technological superpower, a development attributed to the free market reforms initiated in 1991. Coverage of India in the American press, although less extensive, has emphasized the same kinds of facts, with cover stories and op-ed pieces touting the success of the Indian information technology (IT) industry as proof of the benefits of the free market and globalization. Even to those skeptical of the hyperbole about the meteoric rise of India and free market miracles, the stories provide a welcome foil to the standard fare in the Indian media about the myriad failures of the Indian state. For diasporic Indians in the United States and other Western nations, the positive portrayal of India in the Western media has been doubly refreshing, long accustomed as these expatriates are to seeing their country of origin covered in the Western press in terms of Orientalist stereotypes, with mandatory references to the caste system, cows, and poverty. A modicum of national pride, even the most dour cynic would grudgingly admit, is not out of place.

The new millennium, however, has also witnessed the ugly manifestations of misguided national pride in India, exemplified by the events in Gujarat in 2002. On February 27, 2002, a train car of the Sabarmati Express was set afire at Godhra in the western Indian state of Gujarat, allegedly by a mob of Muslims. In the weeks that followed, Hindu nationalist mobs systematically targeted Muslim communities in Gujarat in revenge, setting Muslim-majority villages ablaze and burning Muslims to death. The anti-Muslim violence had the blessings and logistical support of the Hindu nationalist Bharatiya Janata Party (BJP) state government and its chief minister Narendra Modi. The central National Democratic Alliance (NDA) government, a coalition spearheaded by the BJP, reacted slowly to curtail the situation, tantamount to deliberate inaction at a time when an urgent response was the need of the hour.

Neither communal violence nor the dangers of Hindutva, the ideology of Hindu nationalism, are unknown to India. Close to a decade before the Gujarat riots, the country had been afflicted by Hindu-Muslim riots in Mumbai in the state of Maharashtra in December 1992 and January 1993 following the destruction of a mosque by a Hindu nationalist mob on December 6, 1992. On that occasion as well, the state government of Maharashtra had been complicit in the violence against Muslims. But the anti-Muslim pogrom of 2002 was a chilling reminder that the ideology of Hindutva had lost none of its inflammatory and divisive potential in the course of a decade. The events of the year also indicated that foreign direct investment could not be a panacea for all of India's social problems and that Hindu nationalism remained a potent threat to Indian society at a time of economic growth, technological development, and imminent global ascendancy.

The two disparate images of India dominant in media coverage in the last few years can justifiably be seen as a foreshadowing of the fate of the nation in the twenty-first century. What is less obvious is that these strongly contrasting expressions of the Indian state and society should be so deeply and fundamentally related in the imagination of many Indians. Many Hindus in India and elsewhere did not see the actions of Hindu mobs and the state apparatus as anti-Muslim at all. As Ashutosh Varshney points out with regard to the anti-Muslim violence in 2002, "The Hindu right believes that its elected government did exactly what was required: namely, allowing violent Hindu retaliation against the Muslims, including those who had nothing to do with the mob that originally torched the train."[4] For those who held such opinions, there was no necessary contradiction between the brand image of India as an emerging global technological and economic power and its perception as a society at threat of being torn apart by Hindu fundamentalism and nationalism. To the contrary, in the Hindu nationalist perspective, as seen online and offline, in India and in the diaspora, Narendra Modi's actions were viewed as assertions of Hindu *cultural* pride and honor.[5] In tandem, numerous Hindu and Hindu nationalist Web sites strongly emphasize technological and scientific ability as Hindu cultural qualities—another source of Hindu cultural pride and honor.[6]

The juxtaposition of technological achievement with antiminority sentiment as different forms of the same phenomenon—an assertion of cultural and national identity—reflects but one of the modes in which contemporary online Hindu nationalist discourse yokes together technology and nationalism. Other related forms in which the two discourses combine with each other can be seen on Web sites aimed at a global Hindu population. Sites like the comprehensive *Hindu Universe* portal reveal several propositions about Hindu and Indian identity that embody the core claims of technocultural Hindu nationalism.[7]

According to one proposition, Hinduism is a religion that is fully compatible with a technologically and economically interconnected globe. The value of Hinduism as a source of economic tools, skills, or solutions is explained in terms of a foundational cultural universalism purportedly found in the Hindu faith. Web sites like *Hindu Universe* market Hinduism as a globally saleable product. They define the act of economic purchase as a cultural obligation to Hinduism. The activity of consumption, actively encouraged, is itself branded as Hindu. Participation in a chat room discussion about Hindu culture or history on *Hindu Universe*, followed by the act of purchasing a Hindu-branded calling card or cell phone through the same portal, allows an economic transaction to be recast as a more fundamental act of cultural loyalty or duty. The implication is that Hindus and Indians can participate in the world of today *precisely because* they are culturally grounded in a universalist Hindu ethos and heritage that is uniquely attuned to a global marketplace. The post-1991 turn of the Indian state to free market economic policies is tacitly, if not explicitly, recognized as the enabling condition of possibility for this aspect of Hinduism and Hindu culture to flourish in its unhindered form.

Yet another proposition found on the site, squarely in keeping with Hindu nationalist ideology, is that "Indic" identity—the sum of religions, sects, and practices rooted in the history of the indigenous Hindu dharma of the subcontinent—equals and exhausts Indian identity, the former granting the latter its essential character, particularity, and uniqueness: in a word, its soul. Consequently, non-Indic identities, such as those of the Indian Muslim or Indian Christian, do not, strictly speaking, count as Indian in the cultural sense, even though they may qualify as Indian in a legal sense by virtue of constitutional citizenship. This narrative of cultural and national identity privileges its narrow interpretation of Hindu identity as the purest and most authentic incarnation of Indian identity. In addition to the exclusion of Islam and Christianity from the category of Indian faiths, strong condemnation is reserved for Indian Muslims and Christians, as seen in the content on Hindu nationalist sites.

For example, the *Hindu Universe* portal includes a category of content that focuses on the forced conversion of Hindus to Christianity and Islam.[8] The section on Hindu-Muslim relations intones, "Moslem invasion of India has been one of the most brutal events in the history of mankind. During the Moslem [*sic*] rule, millions of Hindus were massacred and converted by force."[9] There are at least two Web sites, *Hindu Holocaust Museum* and *Online Hindu Holocaust Memorial Museum*, dedicated to chronicling the Hindu "holocaust" and "genocide" at the hands of Muslims.[10] The *Hindu Universe* portal also includes a history section, which contains numerous articles espousing anti-Muslim and anti-Christian sentiments.[11] Squarely in keeping with the central premise of Hindutva ideology—the claim that the Indian nation-state is predicated on a Hindu cultural foundation and that Hindu dharma is the overarching and determining structure of identity for all Indians—the articles in the section define Islam and Christianity as alien religions that have been forced on Indian society by Islamic invaders and European colonizers.[12]

In addition to these characteristics, online Hindu nationalism also reflects other tenets of Hindutva ideology. Prime among these is a rejection of the "pseudosecularism" of the Indian state, Congress Party, and Jawaharlal Nehru, the first prime minister of independent India. Other targets are socialism, communism, and Marxism and the individuals, groups, and political parties associated with these ideologies in the Indian context. Left ideologies are deemed responsible for contaminating the academic and intellectual culture of Indian universities, which, the Hindu nationalists argue, is patently anti-Hindu and abusive of Indian heritage and culture. Especially virulent are the attacks aimed at the current president of the Congress Party, Sonia Gandhi.[13]

From the content on Web sites sympathetic to Hindu nationalism, we can identify many of the salient characteristics of technocultural Hindu nationalism. These may be summed up as the following: an insistence on the universalism and globality of Hinduism, a focus on the inherent scientificity of Hindu ethos and culture and the technological expertise of Hindus, ambivalence toward the shifting signifier of the "West," an essentialized interpretation of Hindu and Indian identity accompanied by antiminority prejudice, a rejection of Left ideologies, an embrace of the free market and capitalism, and a rejection of Indian secularism. A look at the profile of the producers and consumers of online Hindu nationalism further enriches our view of the complex entanglements between technology and nationalism as embodied in technocultural Hindu nationalism. There appears to be a significant overlap between the respective profiles of the supporters of Hindu nationalism and Indians working in the technological field in India and overseas. The Internet seems to be the preferred medium and technology of choice for Indian professionals with Hindutva sympathies to promote the ideology and to wage war with competing narratives of national identity.

Following the liberalization of the Indian economy in 1991, IT products and services aimed at global and domestic markets now occupy the pride of place once granted to industrialization as the motor of the Indian economy. Greenspan observes that "inside India, there is no doubt that the IT industry has created jobs, strengthened exports and made substantial contributions to economic growth."[14] India's value for the global economy is perceived as significantly dependent on its IT capabilities. Sachs, Varshney, and Bajpai note that "India is becoming one of the most important players of the world in the IT sector and it is the fastest growing foreign exchange earner for the country."[15] The Internet, in particular, is viewed as both agency and symbol of the benefits of liberalization and technological development. Wolcott and Goodman point out that "the Internet is central to the new vision of India as an IT power in many respects."[16] The importance of the Internet stems from its "core" role in IT development, its potential to profoundly change the nature of business and private communication, its widespread reach, and the fact that it is an agent of change in related "technologies, government policies, laws and services."[17] The Internet is one of the main attractions for American and European companies to set up back-office operations in Indian metros.[18] The Indian central government and numerous Indian state governments seem to agree on the potential of information technology and the Internet for socioeconomic development. Several

Indian states have implemented ambitious e-literacy and e-governance initiatives. National and state government projects are in the process of bringing the benefits of information technology and the Internet to rural Indian communities.

Access to information technology and the Internet, however, is the privilege of an elite in liberalized India. Keniston points out that for "Indians who speak no (or little) English, the barriers to the Information Age are almost insuperable," since fluency in English is a requirement for the use of operating systems, software, and Web sites.[19] The reach of the Internet is far more modest than its instrumental value in driving economic reforms. By February 2005, the total number of phone subscribers in India had increased to 97.03 million.[20] Yet, as of December 2005, India had only 5.56 million Internet subscribers.[21] And the number of Internet users in India as of 2006 was just 40 million—a fraction of its population of more than 1 billion.[22] Internet usage, as Wolcott and Goodman suggest, is dominated by an "urban core" and the predominant users of the Indian Internet are the educated middle classes.[23] It is worth noting here that the Indian "middle classes" are actually socioeconomic elites and can broadly be defined as the top 25 percent income-earning segments of Indian society.[24] Almost 90 percent of all Web sites visited by Indians are in the English language.[25]

The chief beneficiaries of the developments in technology in liberalized India are largely English-speaking middle-class and upper-class urban groups. As will be described shortly, to a significant extent, these are the same demographic segments that have historically benefited from the technology policies of the post-1947 independent Indian state, although there have been some important changes in the internal composition of these elites. Keniston speaks of a new elite group in India comprised of workers in high-tech professions. The "digerati," as Keniston calls them, are "the beneficiaries of the enormously successful information technology (IT) industry and the other knowledge-based sectors of the economy, such as biotechnology and pharmacology."[26] Affluent, cosmopolitan, and globally mobile, the status of the digerati derives from the digerati's technological education and expertise and not from caste, wealth, or social privileges. Through its policies of economic and technological development since the 1980s—but especially the 1990s—the Indian state has also endorsed nonresident Indians (NRIs) working in technology as an immensely valuable resource of human capital for the Indian nation.[27] Muppidi argues that with liberalization the NRI has been inscribed in state discourse as the most authentic incarnation of postcolonial citizenship.[28] Indian expatriates, such as students and professionals, are also a key segment for promoting the growth of Internet usage.[29]

Scholars agree that the core support for the Hindu nationalist movement is drawn from India's middle classes or Indian elites.[30] Corbridge and Harriss view Hindu nationalism as an "elite" revolt, reflecting "the interests and aspirations especially of the middle classes and upper castes."[31] They also note that India's urban socioeconomic elites have provided strong support for the Hindu nationalist BJP in the 1996 and 1999 general elections in India. Urban Indian socioeconomic elites are also among the prime beneficiaries of advances in technology in liberalized India. A strong bastion of support for Hindu nationalism is an overseas, largely U.S.-based, Hindu population.[32] The IT professionals feted by the Indian state are well represented among the diasporic Hindu nationalists.

The Internet seems to be the chosen medium for the Hindu nationalists working in technology to promote Hindutva. "The much touted thousands of computer and software experts," who count among the recent Indian migrants to the United States, "help maintain over five hundred VHP Web sites (www.vhp.org) with their messages of Hindutva, Hindu history, and Muslim-bashing."[33] Rajagopal notes that "with the proliferation of software engineers from India, the internet has become a site for expansion" of Hindu nationalist discourse.[34] According to Vinay Lal, many Hindu nationalist Web sites are produced by politically and culturally conservative Indian software programmers in the United States.[35]

The overlap between the profiles of technology workers and Hindu nationalists suggests the existence of a group whose members view both technological expertise *and* a narrowly defined Hinduness as ideal attributes of Indian identity. The frame of technocultural Hindu nationalism combines these qualities such that technological expertise gets recast in terms of a Hindu cultural ability. As Chakravartty notes, "[C]orporate networks between India and the United States promote a cyber-capitalist rereading of Hindu values, locating the success of high tech Indian entrepreneurs in essential characteristics associated with ethno-religious identity."[36] "For example, Indian competence in the Internet economy," the author observes, "is associated with the 'web of interrelations' that tie together Indian families across national borders."[37] Correspondingly, technocultural Hindu nationalism also rephrases characteristics of cultural identity in the idiom of a model of global significance. For example, on Hindu nationalist Web sites, "Hindu" principles and values are promoted akin to the catchphrases of global corporate capitalism.[38]

This overview provides a sense of the complexities of technocultural Hindu nationalism and the social groups involved in the production, consumption, and propagation of the discourse. But while developments since 1991 allow us to grasp some aspects of the more immediate history of the phenomenon, they do not tell the entire story. It is true that the Indian state has transitioned from a socialist framework to a neoliberal framework within the last two decades, that shift being one of the conditions of possibility and enabling factors for technocultural Hindu nationalism to emerge. But the other factors that influence and shape the discourse—such as the history of Hindu nationalism, the statist valorization of science and technology, the social status accorded to technological skills and occupations—also reflect longer histories and prior incarnations of the relationship between technology and nationalism.

The ideology of Hindu nationalism was formulated as a theory of state and society in the early part of the twentieth century. Marginalized in the decades after independence, it emerged again in visible and conspicuous fashion in the 1980s. Not least among the reasons for its revival were the decaying of Indian political culture under the Congress Party and the accompanying attrition of Congress hegemony in Indian politics and society. Until the 1970s, in the political sphere, the topics of national integration, economic development, and the eradication of poverty and illiteracy were the familiar subjects of electoral campaigns and political agendas. By the 1980s, the terms of political debate had been expanded to accommodate the claims of Hindu nationalism. The vocabulary of Indian politics began to grant a more central role to the idea of India as a Hindu nation and to the necessity for political parties to win over Hindu sentiment. As Brass notes, in the 1984 general elections, "the Hindu majority was mobilized in an atmosphere of hostility to minority demands and behavior. The Congress itself appealed to Hindu nationalism and communalism in this election."[39] The communalization of politics was most visibly reflected in the role played by Indian politicians in polarizing the Hindu and Muslim communities over the controversial Babri Masjid, a disputed religious structure in the city of Ayodhya that both Hindus and Muslims claim as their own. The controversy reflected "the self-conscious and deliberate resurrection by national political leaders of a long dormant local dispute and its transformation into a vital issue affecting the faiths and requiring the solidarity of the two communities."[40] The communalization of Indian politics would bear disastrous consequences; indeed, the political landscape of late twentieth-century India is stained with its bloody legacy. Upadhyaya points out that there were nearly 4,500 incidents of religious violence between 1982 and 1989 "in which over 7,000 people lost their lives, almost four times as many deaths of this type in the 1970s."[41] The riots of 1992 and 1993 that followed the destruction of the Babri Masjid marked, in a sense, the full-fledged return of Hindutva ideology as a formidable political and cultural force in Indian society.

The dominant social understandings of technology and the statist legitimation of scientificity and technological expertise reflected in Hindu technocultural nationalism are also rooted in histories that precede the post-1991 shift. Ironically, they can be seen as partly stemming from the legacy of the socialist, secular vision of Nehru, the very perspective that Hindu nationalists dismiss as anti-Hindu and anti-Indian. Scientificity, socialism, and secularism were the three fundamental components of

Nehruvian nationalism. Together, they embodied the rationalist vision that informed Nehru's dreams and aspirations for independent India. Nehru also strongly emphasized the necessity of cultivating the qualities of scientificity, or the "scientific temper," in Indian society—attributes that embodied a rational enlightened outlook toward the world.[42] In Nehru's vision, the policies of the independent Indian nation-state would reflect the same foundational principle of scientific reason.[43]

In keeping with Nehru's dream, with the onset of independence, the Indian state implemented a program of centralized planning and state ownership of heavy industry. Dams, steel plants, hydroelectric plants, and factories constructed across the country were the new symbols of modern India. Industrial development was given pride of place in the new nation. Over fifty years after independence, the Indian state also set up numerous institutes for industrial training, higher education in engineering and technology, and forty-eight national laboratories for applied scientific research.[44] In independent India, science was authorized as a "reason of state."[45] It would not be an exaggeration to say that science was conferred with a quasi-religious sanctity. Visvanathan notes that "Dams and laboratories were literally the temples of India."[46] The policies of the Nehruvian state also ensured that the institutions of scientific research and development planning were placed outside the political process and immunized from the demands of popular politics.[47] In postcolonial India, a scientific and technological education, especially from the premier educational institutions, operated as a marker of talent and capability and ensured job opportunities and security. Those holding an education or an occupation designated as scientific and technological were perceived as working in the service of the nation, a perception that was internalized and echoed by the Indian people—especially, but not only, socioeconomic elites.

It is one of the profound ironies of Indian history that the technology professionals in India and the United States who vocally advocate the causes of Hindutva and lambaste Nehruvian nationalism owe a significant debt to Nehru's vision of modern India. They are the beneficiaries of the Indian state's socialist policy of subsidizing higher education.

They have benefited from the valorization of scientific education and occupations in independent India. And they have benefited from the notion—strongly endorsed by the Indian state and society—that working in technology-related fields is an act of serving the nation.

But there are other, still older, antecedents of the relationship between technology and nationalism whose shadows still fall on the 1947–1991 context of socialist industrial development and secular nationalism as well as the post-1991 context of neoliberalism and Hindu nationalism. Here, the more distant shores of the colonial past of the Indian nation beckon. Following the conquest of Indian territories by the British East India Company around the mid-eighteenth century, the nascent colonial state in India began the task of accumulating information about the native land and people through topographic, statistical, and other surveys, setting in place a relationship between Western science and technology and the act of defining the Indian nation and people.[48] In the bargain, India was transformed from the heart of darkness to a "space constituted by technics," a land primed for the functioning of roads, railways, and telegraphs.[49] But there were other kinds of consequences, too. For example, Keay suggests that the tremendous human cost of the Great Trigonometrical Survey of India (GTS) and the activities related to the Great Indian Arc of the Meridian of the nineteenth century might have contributed to the tide of anti-British sentiment that culminated in the 1857 Rebellion, the first anticolonial uprising in India.[50]

It was only after 1835, when the colonial administration extended its patronage to scientific education in the medium of English, that Indians were granted the opportunity to prove their scientific abilities.[51] Prakash has argued that the decision to extend scientific learning to Indians embodied a paradox with powerful consequences for a manner of thinking national identity.[52] On the one hand, the decision reaffirmed the discourse of civilizational difference between Europeans and non-Europeans, which defined the European West as rational, enlightened, and civilized and the non-West as primitive and irrational. On the other hand, the decision was an indirect admission by the British that Indians were capable of rational scientific thought. Indians could stake a claim to a universal humanity and rationality via the medium of science, since science embodied universal and rationalist values.[53]

The British decision presented a problematic task for Indian nationalism—the reconciliation of one's universalism with the realities of subjugation under the colonial power.[54] As a solution, a predominantly educated English-speaking Hindu elite, encompassing reformers, historians, writers, and scientists, constructed a narrative of Indianness through a particular interpretation of Hindu oral epics. In this reading, Hinduism was described as a universal scientific, religious, and cultural worldview—a "Hindu science."[55] However, if this science had manifested itself in all its fullness in an ancient time, Indian civilization had been reduced to backwardness because of the savagery inflicted by waves of invaders and the economic exploitation caused by British rule. This recuperation of science as Hindu offered a basis to challenge the hegemony of Western claims over reason, since, the argument went, Hindu thought had been scientific since its very inception, predating the European Enlightenment by centuries.[56]

Scientific knowledge and learning in colonial India operated as an arena in which the supremacy of the West could be contested, yet this contestation meant accepting, to a significant extent, the categories of classification proposed by the colonial state. For, while the Indian elite, through making a case for a "Hindu" science, did not accept the self-definition of science as inherently *Western* in its origins or character, they did, however, accept the self-definition of science as synonymous with rationality and enlightenment. Inasmuch as science stood for the intrinsic cultural superiority of the British colonizers or European civilization, a superiority manifested as a set of capacities that Indians lacked, that proposition was contested by the Indian anticolonial nationalist elites. For example, the Indian anticolonial nationalists articulated a case for the right and capability of Indians to govern themselves on *economic* grounds, arguing that British policies favored European businessmen over Indians, that British rule had resulted in a systematic drain of wealth from India, and that Britain could not reconcile its interests as an imperial power with the economic well-being of Indian society.

Correspondingly, the Indian nationalist elite did not uncritically accept technology as proof of the inherent *cultural* superiority of the West.[57] The imputation of cultural neutrality to technology by the Indian nationalist elite meant paradoxically that they could adopt it as Indian without the fear that such adoption would have a negative cultural impact on Indian communities.

It is not too difficult to see the debt owed, respectively, by the Nehruvian and the Hindu nationalist perspectives to the Indian anticolonial nationalist elite's understanding of the relationship between technology and nationalism. The Nehruvian framework took the obsession with scientificity as an embodiment of national identity and incorporated it into a socialist and secular nationalist definition of the Indian nation-state and society. The Hindu nationalist framework, as seen online, combines the statist emphasis on scientificity with anticolonial nationalist arguments about Hindu science and Hindu revival into an exclusivist definition of national identity. It also reworks these ideas into its projections of global ascendancy for the Hindu-Indian nation. In all three historical phases—the anticolonial nationalist, Nehruvian socialist, and neoliberal Hindu nationalist—science and technology operate as signifiers of service to the nation and markers of social and cultural status in Indian society. Similarly, what has endured across the anticolonial nationalist, Nehruvian, and Hindu nationalist paradigms is the multiple, if paradoxical, symbolizations of science and technology as simultaneously, or alternately, Western, universal, and Indian.

In each historical phase, the associations between technology and nationalism have condensed in ideas about self and other; they have been incorporated in imaginings of the state and the nation; and they have materialized as claims about identity, community, and society. From the nineteenth century onwards, Hindu elites have also defined scientificity as a Hindu quality to be cultivated within the individual and the cultural group. Technology, on the one hand, has enabled the conflation of Indian and Hindu identity. On the other hand, it has also stood for a transcending of particularist identities as a means of realizing a pan-Indian universal identity. In either formulation, it has granted the possessor of technological skills social status both as an individual and as a member of a larger imagined national community. This is the history that the book engages with, en route to explicating the characteristics of technocultural Hindu nationalism. The story allows one to read technocultural Hindu nationalism as a palimpsest of the many overlapping and contradictory histories that, like tributaries, have flowed into its making. The reading undertaken here also seeks to unpack the complex collusions and mutual reinforcements across discourses of state and society in affirming the associations between nationalism and technology in each historical era.

The characteristics of technocultural Hindu nationalism arise from a complex genealogy, with its attendant twists and turns and continuities and discontinuities over the course of historical time. But the discourse of technocultural Hindu nationalism is also shaped by its *online* location and the context of its production and consumption in cyberspace. The question at stake may be expressed as follows: what is the consequence of the intersection of Hindu nationalist discourse and the Internet? A comparison with other Indian nationalisms online appears to be one useful framework to employ in seeking to answer the question. To be more specific, a comparative analysis of the relative distribution of themes, such as scientificity and globality, across different Indian nationalisms online can help establish which themes are foregrounded as a result of the medium of the Internet and which are prominent as a result of their centrality to the particular nationalist discourse. And, relatedly, the impact of the modalities of Internet communication on the forms taken by Hindu nationalism in cyberspace, its similarities and differences from its "offline" avatars, also demands attention.

My analysis takes shape as an interdisciplinary endeavor, combining qualitative and quantitative research methodologies and drawing on historical scholarship about South Asia, social and cultural theory, and the sociology of new media—specifically, the evolving field of Internet studies. The theoretical framework utilizes key concepts from the work of sociologist Pierre Bourdieu, specifically the concepts of class, field, and capital, and the theory of the state. Two important characteristics of Bourdieu's framework, which indicate the mode in which it is used in this work, warrant noting here. First, Bourdieu's poststructuralist project does not propose an ontology of the social but is primarily descriptive. Second, Bourdieu's reservoir of theoretical concepts constitutes a "metatheory," which allows it to be used *with* other theoretical frameworks, such as the insights of the Subaltern Studies school of South Asian historiography, the Gramscian notion of "legitimation crisis," or the concept of the "network society."[58] In this regard, Brubaker notes the "extreme generality" of Bourdieu's theoretical conceptions, while Sulkunnen speaks of Bourdieu's flexible and "plastic" sociology.[59] Bourdieu's example in his book *Distinction* of undertaking quantitative sociological analysis through the exploratory statistical technique of correspondence analysis also presents a viable method for mapping the distribution of themes in Indian nationalisms online.[60]

Chapter 2 presents the theoretical and methodological framework employed in the study. Critical issues that warrant theoretical reorientation of the concepts in reading Indian history are also examined here. These are (a) the problematic of instrumentality in Bourdieu's theory, (b) the problematic of Eurocentricity, and (c) the problematic of the religious field as theorized by Bourdieu. Bourdieu's theory has been critiqued on the grounds that it is excessively deterministic and totalizing—for example, offering no space for political resistance—which also exposes him to the criticism that he does not satisfactorily explain how his analysis is exempt from the logic of reproducing dominant power structures.[61] This charge may be answered in part by pointing to the descriptive and metatheoretical nature of Bourdieu's framework, which does not necessarily preclude the possibility of agency or political action but seeks to demonstrate how such possibilities themselves are intertwined with more fundamental logics of social practice. Additionally, Bourdieu draws attention to the fact that the analytical categories that he employs, such as "classes," are theoretical constructs. Social space, as Bourdieu describes it, is composed of a network of *possible* contestations and alliances between groups across different fields and not of entities that correspond to empirical fixities. Combined with Bourdieu's emphasis on reflexive sociology, this approach illuminates, at once, the critical explicatory value of the framework *and* its limits. In the case of colonial and postcolonial Indian society, for instance, the limits are revealed with regard to the formulation of the religious field, the question of Eurocentricity, and the problematic of instrumentality. Inasmuch as there is a risk of a totalizing or mechanistic application of theory to context—which holds for any theoretical framework and not just that of Bourdieu—the use of other key theoretical ideas or concepts enhances the heuristic power of the framework, as well as the possibility of internal critique. Bourdieu's framework, then, does not claim to be exempt from the play and logic of power structures, but vests its truth claims in a rigorously wrested autonomy that is necessary for social scientific inquiry.

Chapter 3 illustrates the Janus-faced roles of science and technology during the early period of colonial rule and anticolonial nationalism. The chapter covers the introduction of Western science and technology in India and focuses on interventions of the colonial state in educational, cultural, and other domains as it consolidated its presence in the Indian subcontinent. As instruments of the colonial state, Western science and technology played a key role in the upheavals wrought by the rupture of colonialism upon indigenous political traditions, economic structures, and modes of sociality in India. However, Western science and technology were also reconfigured and claimed as instruments of national empowerment by an English-speaking, educated Hindu elite. In elite Indian discourse, scientific expertise was constructed as a traditional Hindu attribute and the means to a unique Indian modernity. This conception of scientificity was inextricably woven into emergent narratives of anticolonial nationalism. The polyvalent status ascribed at this originary moment to science and technology in India—as Western and Eastern, national and universal, Hindu and Indian—set in place a template that recurs in later historical periods.

Chapter 4 traces the development of the technological field and capital in India over the course of much of the twentieth century. I approach the relationship between technology and nationalism in the phase prior to independence in 1947 through an assessment of three competing visions for the Indian nation-state, each of which offers a philosophical perspective about the nature of the state and proposes a corresponding role for technological development as a matter of state responsibility. The models evaluated are Gandhi's dream of an Indian nation of villages, premised on a radical critique of Western industrial modernity; the Hindu state theorized by V. D. Savarkar, the pioneering ideologue of Hindu nationalism; and the socialist scheme of an industrially advanced and socially progressive India proposed by Jawaharlal Nehru. The chapter also covers what may be called the era of Nehruvian socialism (1947–1991), given the triumph of Nehru's vision over the other models at the moment of independence. In Nehruvian India, science and technology were expected to cure the nation of its economic and social backwardness. The idea of the scientific temperament as a highly desirable national characteristic was also strongly reinforced through state policies. I describe how Indian socioeconomic elites interpreted this statist discourse, accepting the state's endorsement of scientific and technological capital but not necessarily the idea that science should serve the objectives of social justice.

Chapter 5 follows the relationship between technology and nationalism since 1991, with reference to the liberalization and globalization of the Indian economy as well as the resurgence of Hindu nationalism since the 1980s. Each of these paradigms proposes a particular definition of the Indian nation and a blueprint for national progress that will help realize India's potential in the globalized world of the twenty-first century. The convergence between the two paradigms reflects a new form of the relationship between technology and nationalism—technocultural Hindu nationalism—typically seen in Web sites addressed to a global Hindu audience.

I analyze the characteristics of technocultural Hindu nationalism with reference to wider historical, social, cultural, and economic contextual factors. I argue that the claim presented on Hindu nationalist Web sites that scientificity is a simultaneously Hindu, Indian, and global trait is strongly rooted in early Indian anticolonial nationalist consciousness. Similarly, technocultural Hindu nationalism draws significantly on the legacy of the statist logic of technological development articulated in

Nehruvian India, even though Hindu nationalism opposes Nehruvian notions of socialism and secularism. Technocultural Hindu nationalism, while indebted to Savarkar's theorization of the Hindu state, also refashions Savarkar's views to reflect the interests of an affluent diasporic Hindu population and the demands of global capitalist modernity.

Chapter 6 presents an exploratory quantitative analysis of the characteristics of online Indian nationalisms. Using the technique of correspondence analysis, I undertake a comparative statistical mapping of the distribution of key themes, including scientificity, globality, and rights and freedoms, across Hindu, Sikh, and Kashmiri nationalist Web sites. Each theme has been selected on the basis of its relevance for nationalism as well as for *online* nationalism. The rationale for the selection of the theme of scientificity is self-evident, given its centrality to Hindu nationalist discourse. With regard to the theme of globality, for instance, the argument is that nationalisms will market themselves in cyberspace as global issues—of relevance to the international community at large, as well as for diasporic ethnic, cultural, or religious groups dispersed across the world—to speak to the potentially global audience using the Internet. The theme of rights and freedoms is included based on the reasoning that a nationalist movement will typically emphasize both the political right to sovereignty and other political, cultural, social and economic rights, with reference to domestic constitutional and legal frameworks as well as the framework of international human rights. The central question explored in the chapter is whether the core characteristics of technocultural Hindu nationalism, such as its emphasis on scientificity, are more reflective of the medium of the Internet than of the substantive content of the ideology. My argument is that the emphasis on scientificity is peculiar to the history of Hindu nationalism and not an artifact of the virtual or online status of the discourse. Accordingly, the primary hypothesis tested in the chapter is that the theme of scientificity should appear more prominently in Hindu nationalist Web sites than in other similar types of Web sites. The secondary hypothesis tested is that the theme of scientificity should occur prominently in Hindu nationalist Web sites at large across the World Wide Web.

In the concluding chapter, I reflect on the possibilities offered by the Internet for reimagining Indian identities in light of the historical associations between nationalism and technology. I extend the discussion of Indian identities online to the complexities of the production, representation, and consumption of online discourse in general. I outline some directions for theorizing the relationship between the modalities of Internet communication and online expressions of identity. I also offer a reflection on the viability of the theoretical framework employed in the book, based on the reading and analysis in earlier chapters.

The role of technology in imaginings of the Indian nation and in ideological projects of nation building in the colonial era and the Nehruvian era has been assessed in some depth.[62] There is a large body of work on various aspects of Hindu nationalism.[63] The use of the Internet for the promotion of Hindu nationalist ideologies has also received some attention from scholars, typically in essays or articles that address contemporary expressions of the ideology.[64] However, the relationship between technology and Hindu nationalism as it takes shape on the Internet has not been assessed in sufficient depth as a phenomenon *in and of itself*. This phenomenon, which I have termed "technocultural Hindu nationalism," has not been sufficiently historicized with respect to the longer history of technology and nationalism in Indian society, specifically, in terms of how that longer history, with its shifts over time, influences the character of the discourse.[65] For example, while Hindu nationalism is opposed to the Nehruvian understanding of India as a multicultural and multireligious society, technocultural Hindu nationalism's obsession with scientificity is *also* strongly influenced by the Nehruvian emphasis on scientific and technological development. The ambivalence of technocultural Hindu nationalism with regard to Nehru's legacy has not been adequately addressed in analyses of online Hindu nationalism either. The analysis of technocultural Hindu nationalism offered in this book addresses such complexities of the discourse.

Studies of Internet discourse are often characterized by insufficiently theorized notions of online communities as well as by either romanticized or neo-Luddite conceptions of technology.[66] In relating the historical contextualization of online discourse to a sociological analysis of Internet usage, this book calls for a rethinking of assumptions about virtual communities, cyberspace, and practices of Internet use. Finally, this book contributes to debates about Bourdieu's concepts of field, class, and capital, and the theory of the state.

Of all the grand narratives that suffuse the space of public and private life in contemporary India, none is as visible and manifest as the discourse of the nation and nationalism. That discourse, no doubt, is deeply contested; it occurs as an amalgam of competing secular, secessionist, and majoritarian claims about the Indian state and people. Nonetheless, the *idea* of the nation, whatever its many and contested meanings, is ubiquitous in Indian social space. It is found in political slogans on billboards and on the sides of buses; in the coverage of the Indian cricket team; in the stories and melodies of the Hindi film industry; in the offerings of the Indian advertising industry; in the speeches of the muezzin, pandit, and priest declaimed from mosque, temple, and church; in the bodily practices of members of the Hindu nationalist Rashtriya Swayamsevak Sangh; and in the wars between Indian mafia gangs from Mumbai to Uttar Pradesh.

Perhaps only marginally less prominent in Indian social space, especially in the present historical context, is the discourse of technology. In the nation's metros and small towns, the *idea* of technology is all-pervasive, although, ironically, a significant proportion of the population in India does not have access to the benefits and conveniences of technology. The pervasiveness of technology can be seen in the cybercafés that keep proliferating across Indian cities, in the mushrooming of dozens of private educational institutions that offer short-term diplomas in IT and computing skills, in advertisements that lure students with a promise that they will get jobs as software programmers in America, and in the media stories that sing the glories of Indian technological achievements. Less visible but no less tangible evidence of the pervasiveness of the discourse can be found in the aspiration of hundreds of thousands of high school students to obtain a seat in one of India's premier technology institutes. It can also be seen in the annual migration of thousands of Indian graduates in science and technology to the United States for higher studies. The emphasis on technology is also bolstered by the numerous announcements of initiatives in technological development by the Indian state and private sector, duly noted and reported by the Indian press.

The omnipresence of nationalism and technology as two intertwined master narratives of Indian social life reflects the burdens and legacies of Indian colonial and postcolonial modernity. The relationship manifests itself in the present in one form in the shape of what I have termed "technocultural Hindu nationalism," embodied in the figure of the Hindu who seeks to redefine the meaning of Indianness on and through the Internet. One hesitates from speculating too much about the possible future trajectories of the relationship between technology and nationalism in India. The largely descriptive analysis I have undertaken in this book precludes such predictions about the future. One hopes, however, that this relationship in the future—whether within or without the realm of cyberspace—takes the form of a happy and democratic profusion of identities that are inclusive and enabling rather than exclusive and constricting. Otherwise, the idea of India as a land of pluralism, tolerance, and coexistence may be reduced to no more than an article of faith in the years to come, even as the nation prepares to don its new mantle as an economic and technological powerhouse in the twenty-first century.

ENDNOTES

1. Jawaharlal Nehru, "An Address at the Indian Science Congress, Calcutta, 26 December 1937," speech documented in *Jawaharlal Nehru: An Anthology*, ed. Sarvepalli Gopal (New Delhi: Oxford University Press, 1983), 442.
2. Gilles Deleuze and Félix Guattari, *Nomadology: The War Machine*, trans. Brian Massumi (New York: Semiotext(e), 1986), 65.
3. Gyan Prakash, *Another Reason: Science and the Imagination of Modern India* (Princeton: Princeton University Press, 1999). *See esp*. 8–9, 46–48, and 91.
4. Ashutosh Varshney, "Understanding Gujarat Violence," *Contemporary Conflicts*, March 26, 2004, http://conconflicts.ssrc.org/gujarat/varshney/ (accessed September 23, 2008). In fact, Narendra Modi, the chief minister of the Gujarat state government that presided over the anti-Muslim violence, was invited to speak at a conference in New York scheduled for March 2005, organized by the Asian American Hotel Owners Association. A campaign by the Coalition Against Genocide, a grouping of Indian organizations opposed to Modi's visit, was successful in persuading the U.S. government to deny Modi a visa to visit the United States. The decision was met by outrage among many Hindus in the United States. For a description of the reaction of Indians in the United States to the decision, see Vijay Prashad, "No Entry for Modi," *Frontline* 22, no. 7, March 12–25, 2005, http://www.hinduonnet.com/fline/fl2207/stories/20050408003313100.htm (accessed March 12, 2006).
5. Those who invited Modi for the New York event are part of the "Hindu Diaspora, mostly Gujaratis who have been vociferously praising the role of Modi during the riots in Gujarat in February March–April 2002." M. A. Rane, "Narendra Modi Denied Entry to the United States," *PUCL Bulletin*, August 2005, http://www.pucl.org/Topics/Religion-communalism/2005/modi-usa.htm (accessed March 19, 2006). For Modi's rhetoric of Gujarati and Hindu pride, see "The Modi Factor," *The Hindu*, June 27, 2004, http://www.hindu.com/2004/06/27/stories/2004062700731400.htm (accessed March 19, 2006).
6. See, for example, Sudheer Birodkar, *Ancient India's Contribution to Our World's Material (Temporal) Culture*, n.d., http://www.geocities.com/Lavlesh/sudheer_contributions/index.html (accessed March 19, 2006); Global Hindu Electronic Networks, "Sciences," *Hindu Universe: Hindu Resource Center*, n.d., http://www.hindunet.org/sciences/ (accessed September 17, 2008); Himalayan Academy, "In Science and Beyond—*Vijnane*

Paratashcha," in *Loving Ganesha: Hinduism's Endearing Elephant-Faced God*, 2nd ed. (United States: Himalayan Academy, 2000), chap. 4, http://www.himalayanacademy.com/resources/books/lg/lg_ch-04.html (accessed March 19, 2006); and Sushama Londhe, "Advanced Concepts of Hinduism," *A Tribute to Hinduism*, updated September 7, 2005, http://www.atributetohinduism.com/Advanced_Concepts.htm (accessed March 19, 2006).

Many Web sites sympathetic to Hindu nationalism include a section on science and technology. These sites strongly emphasize the scientificity of Hinduism and Hindu culture through a range of claims. It is often claimed, for instance, that ancient Hindu civilization and culture anticipated the major findings of modern science. Not all the Hindu Web sites that present such claims appear to endorse Hindu nationalist ideology in unequivocal terms. Nonetheless, the majority of sites appear to share the basic assumption of Hindu nationalist ideology that Indian civilization is synonymous with Hindu civilization and that Hindu culture defines Indian culture.

7. Global Hindu Electronic Networks, *Hindu Universe: Hindu Resource Center*, n.d., http://www.hindunet.com (accessed February 19, 2005).
8. Global Hindu Electronic Networks, "Conversions," *Hindu Universe: Hindu Resource Center*, n.d., http://conversions.hindunet.org/ (accessed February 12, 2005).
9. Global Hindu Electronic Networks, "Relationship Between Hindu And Moslem," *Hindu Universe: Hindu Resource Center*, http://www.hindunet.org/hindu_moslem/ (accessed February 12, 2005).
10. *Hindu Holocaust Museum*, n.d., http://www.mantra.com/holocaust/ (accessed July 2, 2005) and *Online Hindu Holocaust Museum*, n.d., http://sarvadharma.org/Museum/HinduHolocaustMuseum.htm (accessed July 2, 2005).
11. Global Hindu Electronic Networks, "History," *Hindu Universe: Hindu Resource Center*, n.d., http://www.hindunet.org/hindu_history/hindu_history.html (accessed February 12, 2005). Several Web pages under the History section of the portal refer to the section as the "Library of Hindu History," suggesting that the section can also be considered a self-contained Web site.
12. See, for instance, Global Hindu Electronic Networks, "In the Islamic Ages," *Hindu Universe: Hindu Resource Center*, n.d., http://www.hindunet.org/hindu_history/modern/moghal_link.html (accessed February 12, 2005).
13. For the Hindu nationalist perspective on secularism, see, for instance, Ramen Bando, *Secularism Extra-Ordinary* (London: Sahitya Porishad, 1997), http://www.hvk.org/specialrepo/seo/index.html (accessed March 19, 2006). The book is accessible through the Web site of a Hindu nationalist orga-

nization, Hindu Vivek Kendra, dedicated to the promotion of Hindutva. See *Hindu Vivek Kendra*, n.d., http://www.hvk.org/ (accessed March 19, 2006). See also the blog *Pseudo-Secularism*, n.d., http://pseudosecularism.blogspot.com/ (accessed March 19, 2006).

For the Hindu nationalist view of Marxist intellectuals, see, for example, Koenraad Elst, "The Politics of the Aryan Invasion Debate," *The Koenraad Elst Site*, April 2003, http://koenraadelst.bharatvani.org/articles/aid/aryanpolitics.html (accessed March 18, 2006).

When the distinguished Indian historian Romila Thapar, a principled critic of Hindu nationalism, was appointed as the first holder of the Kluge Chair in Countries and Cultures of the South by the Library of Congress, Hindu nationalists in the United States circulated a petition on the Internet to protest the appointment. The petition described Thapar's appointment as a waste of "American resources on a Marxist idealogical [*sic*] assault on Hindu civilization." See "Protest US Supported Marxist Assault against Hindus," *Petition Online*, n.d., http://www.petitiononline.com/108india/petition.html (accessed March 19, 2006).

For a description of Hindu nationalist views on Sonia Gandhi, see Praful Bidwai, "Plebiscite on Prejudice?" *Frontline* 16, no. 19, September 11–19, 1999, http://www.hinduonnet.com/fline/fl1619/16191030.htm (accessed March 19, 2006). As an example of a Hindu nationalist Web site about Sonia Gandhi, see *Sonia Maino, Sonia Gandhi*, n.d., http://www.soniamaino.com/ (accessed March 20, 2006). On the home page, the site self-identifies as a "Hindu Indian nationalist blog." See also, "About Us," *Hindu Unity*, n.d., http://hinduunity.org/aboutus.html (accessed March 19, 2006).

14. Anna Greenspan, *India and the IT Revolution: Networks of Global Culture* (New York: Palgrave, 2004), 109.
15. J. D. Sachs, A. Varshney, and N. Bajpai, "Preface to the Paperback Edition," in *India in the Era of Economic Reforms*, ed. J. D. Sachs, A. Varshney, and N. Bajpai (New Delhi: Oxford University Press, 2000), xi.
16. Peter Wolcott and Seymour Goodman, "Global Diffusion of the Internet I: Is the Elephant Learning to Dance?" *Communications of the Association for Information Systems* 11, no. 32 (2003): 565, http://cais.isworld.org/articles/11-32/default.asp?View=Journal&x=34&y=14 (accessed May 17, 2005).
17. Ibid., 566.
18. Sachs et al., "Preface to the Paperback Edition," xi–xii.
19. Kenneth Keniston, "Introduction: The Four Digital Divides," in *IT Experience in India: Bridging the Digital Divide*, ed. Kenneth Keniston and Deepak Kumar (New Delhi: Sage Publications, 2004), 15.

20. "Number of Subscribers in Cellular (Including WLL) and Fixed Line in India (2000 to 2005)," *Indiastat.com*, n.d., http://www.indiastat.com/india/ShowData.asp?secid=368487&ptid=19279&level=4 (accessed January 10, 2006).
21. "Number of Internet Subscribers in India (as on 13.03.1999 to 30.06.2005)," *IndiaStat.com*, n.d., http://www.indiastat.com/india/ShowData.asp?secid=395675&ptid=143&level=3 (accessed June 26, 2007).
22. "India: Internet Usage Stats and Telecommunications Market Report," *Internet World Stats*, n.d., http://www.internetworldstats.com/asia/in.htm (accessed June 19, 2007).
23. Wolcott and Goodman, "Global Diffusion of the Internet I," 612 (accessed May 17, 2005).
24. Achin Vanaik, "Consumerism and New Classes in India," in *Thinking Social Science in India: Essays in Honour of Alice Thorner*, ed. Sujata Patel, Jasordhara Bagchi, and Krishna Raj (New Delhi: Sage Publications, 2002), 228. According to different parameters, the elite Indian "middle classes" number anywhere between 100 and 250 million (ibid.).
25. Wolcott and Goodman, "Global Diffusion of the Internet I," 612 (accessed May 17, 2005).
26. Keniston, "Introduction: The Four Digital Divides," 17.
27. Paula Chakravartty, "The Emigration of High-Skilled Indian Workers to the United States: Flexible Citizenship and India's Information Economy," in *The International Migration of the Highly Skilled: Demand Supply, and Development Consequences in Sending and Receiving Countries*, ed. Wayne A. Cornelius, Thomas J. Espenshade, and Idean Salehyan (La Jolla, CA: Center for Comparative Immigration Studies, 2001), 327.
28. Himadeep Muppidi, *The Politics of the Global* (Minneapolis: University of Minnesota Press, 2004), 56.
29. Wolcott and Goodman, "Global Diffusion of the Internet I," 613 (accessed May 17, 2005).
30. Stuart Corbridge and John Harriss, *Reinventing India: Liberalization, Hindu Nationalism and Popular Democracy* (Malden, MA: Blackwell, 2000); Thomas Blom Hansen, *The Saffron Wave: Democracy and Hindu Nationalism in Modern India* (Princeton: Princeton University Press, 1999); and Vanaik, "Consumerism and New Classes in India," 227–234.
31. Corbridge and Harriss, *Reinventing India*, xix.
32. Arvind N. Das, "The End of Geography: Nationalism in the Era of Globalization," in *Nations Under Siege: Globalization and Nationalism in Asia*, ed. Roy Starr (New York: Palgrave, 2002), 31–62; Vinay Lal, "The Politics of History on the Internet: Cyber-Diasporic Hinduism and the North

American Hindu Diaspora," in *In Diaspora: Theories, Histories, Texts*, ed. M. R. Paranjpe (New Delhi: Indialog Publications, 2001), 179–221; Sucheta Mazumdar, "The Politics of Religion and National Origin: Rediscovering Hindu Indian Identity in the United States," in *Antinomies of Modernity: Essays on Race, Orient, Nation*, ed. Vasant Kaiwar and Sucheta Mazumdar (Durham: Duke University Press, 2003), 223–260; and Arvind Rajagopal, "Hindu Nationalism in the US: Changing Configurations of Political Practice," *Ethnic and Racial Studies* 23, no. 3 (May 2000): 467–496.

33. Mazumdar "The Politics of Religion and National Origin," 240.
34. Rajagopal, "Hindu Nationalism in the US," 484.
35. Vinay Lal, "The Politics of History on the Internet," 179.
36. Chakravartty, "The Emigration of High-Skilled Indian Workers to the United States," 343.
37. Ibid.
38. For example, the Hindu Students Council, whose Web site is part of the Global Hindu Electronic Networks, is defined as a "voluntary run organization committed to realizing the ancient Vedanta truths such as *Vasudaiva Katumbakum* (The Whole World is One Family)." See Global Hindu Electronic Networks, "About This Site," *Hindu Universe: Hindu Resource Center*, n.d., http://www.hindunet.com/contact.htm (accessed February 19, 2005). The language on the Hindu Students Council's Web site, with its corporatist inflections, seems especially suited for its target demographic of students and professionals. The "About HSC" section of the Web site states,

> HSC represents the first and only North American and international attempt by students and young professionals like yourself to explore, discover & experience the immense treasures of the time-tested knowledge and wisdom of the great Hindu culture. Indeed, we have within our grasp the vast philosophical, spiritual, social and scientific tools of the Hindu System. Let us learn to use them for achieving excellence in personal life and to enrich our community, our environment and our world.

See "About HSC," *Hindu Students Council*, n.d., http://www.hscnet.org/aboutus.php (accessed January 4, 2006).

When viewed on September 4, 2008, the content on the Global Hindu Electronic Networks "About This Site" page had changed with no mention of the relationship between the Global Hindu Electronic Networks and the Hindu Students Council. This appears to be an outcome of one of the initiatives undertaken by the Campaign to Stop Funding Hate (CSFH), a

U.S.-based organization dedicated to countering the spread of Hindutva ideology in India and in the United States. The CSFH report, "Unmistakably Sangh: The National HSC and Its Hindutva Agenda," points to the relationship between Hindu nationalist organizations and the Hindu Students Council. See Campaign to Stop Funding Hate, "Unmistakably Sangh: The National HSC and Its Hindutva Agenda," *Campaign to Stop Funding Hate*, n.d., http://hsctruthout.stopfundinghate.org/ (accessed September 4, 2008). One of the press releases on the CSFH Web site notes that the Hindu Students Council has sought to conceal its relationship with Hindu nationalist organizations by changing the contact page on the *Hindunet* Web site (which is the same as the *Hindu Universe* Web site). See Campaign to Stop Funding Hate, "The Cover-Up Begins: National HSC Goes One Step Further to Dupe its Membership," *Campaign to Stop Funding Hate*, n.d., http://hsctruthout.stopfundinghate.org/PRs/pr_Question2.html (accessed September 4, 2008).

39. Paul R. Brass, *The Politics of India since Independence*, 2nd ed., *The New Cambridge History of India*, part 4, vol. 1 (Cambridge, U.K.: Cambridge University Press, 1994), 238.
40. Ibid., 239–241. For a detailed assessment of the communalization of Indian politics since the 1980s, see chapter 7 ("Communal and Caste Conflict: Secularism, Hindu Nationalism and the Indian State"), 228–266.
41. Prakash Chandra Upadhyaya, "The Politics of Indian Secularism," *Modern Asian Studies* 26, no. 4 (October 1992): 821–822.
42. Jawaharlal Nehru, *The Discovery of India* (Calcutta: Signet Press, 1946; New Delhi: Jawaharlal Nehru Memorial Fund and Oxford University Press, 1995), 512 (page citations are to the reprint edition).
43. See Partha Chatterjee, "Development Planning and the Indian State," in *The State, Development Planning and Liberalisation in India*, ed. Terence J. Byres (New Delhi: Oxford University Press, 1997), 82–103; Sunil Khilnani, *The Idea of India* (New Delhi: Penguin, 1999), 61–88; Ashis Nandy, "Introduction: Science as a Reason of State," in *Science, Hegemony and Violence: A Requiem for Modernity*, ed. Ashis Nandy (Tokyo: United Nations University, 1988), 1–23; and Shiv Visvanathan, *A Carnival for Science: Essays on Science, Technology, and Development* (New Delhi: Oxford University Press, 1997), 4–5.
44. See chapter 3, "Industrial Development & Progress after Independence," in Government of India, Ministry of Labour, *Main Report*, vol. 1, *Report of the Second National Commission on Labour*, 2002, http://www.labour.nic.in/lcomm2/2nlc-pdfs/Chap3.pdf (accessed May 11, 2004).

45. Nandy, "Introduction," 1.
46. Visvanathan, *A Carnival for Science*, 4.
47. See Chatterjee, "Development Planning and the Indian State," 82–103, and Nandy, "Introduction," 1–23, for a discussion of the privileging of science over politics in Nehruvian India.
48. Zaheer Baber, *The Science of Empire: Scientific Knowledge, Civilization, and Colonial Rule in India* (Albany: State University of New York Press, 1996), 137.
49. Prakash, *Another Reason*, 11.
50. John Keay, *The Great Arc: The Dramatic Tale of How India Was Mapped and Everest Was Named* (London: HarperCollins, 2000). *See esp*. 169–170.
51. Baber, *The Science of Empire*, 138.
52. Prakash, *Another Reason*, 8–9, 46–48.
53. Ibid., 108.
54. Ibid., 8–9. See also 98–120.
55. Ibid., 9.
56. Ibid., 84. For our purposes, it is the ideological mobilization of the *idea* of Indian scientificity as the basis of anticolonial nationalism that is of interest. However, the claim of the Hindu intelligentsia is partly true. In *The Science of Empire*, Baber describes ancient Indian innovations in mathematics, astronomy, and medicine in some detail. See chapter 2, "Science, Technology, and Social Structure in Ancient India," 14–52.
57. See Partha Chatterjee, *The Nation and Its Fragments: Colonial and Postcolonial Histories* (Princeton: Princeton University Press, 1993), 6. Chatterjee presents a compelling case for the autonomy of an Indian cultural field from the framework of state reason in the colonial era. Chatterjee argues that anticolonial nationalism visualizes an autonomous "domain of sovereignty within colonial society well before it begins its political battle with the imperial power" (ibid.). He describes this as a separation between a realm of cultural or spiritual values, an "inner" domain where those colonized hold authority, and the "domain of the outside," the sphere of "economy and statecraft, of science and technology," where the West has vanquished the natives (ibid.). Indian anticolonial nationalism, accordingly, defined its task as one of wresting back control of the outer domain, while it could rest assured that its cultural identity was uncontaminated by the colonial encounter.
58. Rogers Brubaker, "Rethinking Classical Theory: The Sociological Vision of Pierre Bourdieu," in *Pierre Bourdieu*, ed. Derek Robbins (London: Sage, 2000), 3: 91.
59. Ibid., 109; Pekka Sulkunnen, "Society Made Visible: On the Cultural Sociology of Pierre Bourdieu," in *Pierre Bourdieu*, ed. Derek Robbins (London: Sage, 2000), 3: 223.

60. Pierre Bourdieu, *Distinction: A Social Critique of the Judgement of Taste*, trans. Richard Nice (Cambridge, MA: Harvard University Press, 1984).
61. See, for example, John Frow, "Accounting for Tastes: Some Problems in Bourdieu's Sociology of Culture," in *Pierre Bourdieu*, ed. Derek Robbins (London: Sage, 2000), 3: 44–58. In Frow's words, it is not clear "what constitutes the possibility of Bourdieu's own critical exteriority to the dominant norms" (57).
62. See, for example, Baber, *The Science of Empire*; Ashis Nandy, "Introduction," 1–23; Prakash, *Another Reason*; Satpal Sangwan, *Science, Technology and Colonisation: An Indian Experience 1757–1857* (New Delhi: Anamika Prakashan, 1991); Visvanathan, *A Carnival for Science*; and Visvanathan, *Organizing for Science: The Making of an Industrial Laboratory* (New Delhi: Oxford University Press, 1985).
63. See, for example, Rajeev Bhargava, "The Cultural Nationalism of the New Hindu," *Dissent* (Fall 2003): 11–17; Chetan Bhatt, *Hindu Nationalism: Origins, Ideologies and Modern Myths* (New York: Berg, 2001); Sumantra Bose, "'Hindu Nationalism' and the Crisis of the Indian State: A Theoretical Perspective," in *Nationalism, Democracy and Development: State and Politics*, ed. Sugata Bose and Ayesha Jalal (New Delhi: Oxford University Press, 1997), 104–164; Corbridge and Harriss, *Reinventing India*; Hansen, *The Saffron Wave*; Meera Nanda, *Prophets Facing Backward: Postmodern Critiques of Science and Hindu Nationalism in India* (New Brunswick: Rutgers University Press, 2003); Ronald Neufeldt, "The Hindu Mahasabha and Gandhi," in *Indian Critiques of Gandhi*, ed. Harold Coward (Albany: State University of New York Press, 2003), 131–152; Bindu Puri, "Indian Nationalism and the Hindu Argument: Savarkar and Gandhi," in *Mahatma Gandhi and His Contemporaries*, ed. Bindu Puri (Shimla, India: Indian Institute of Advanced Study, 2001), 154–176; Tanika Sarkar, "Heroic Women, Mother Goddesses: Family and Organisation in Hindutva Politics," in *Women and Right-Wing Movements: Indian Experiences*, ed. Tanika Sarkar and Urvashi Butalia (London: Zed Books, 1995), 181–215; and Achin Vanaik, *The Furies of Indian Communalism: Religion, Modernity, and Secularization* (London: Verso, 1997).
64. See, for example, Das, "The End of Geography," 31–62; Vinay Lal, "Aryavarta and Silicon Valley: Indian History on the Net," in his *The History of History: Politics and Scholarship in Modern India* (New Delhi: Oxford University Press, 2003), 231–273; Vinay Lal, "The Politics of History on the Internet," 179–221; Mazumdar, "The Politics of Religion and National Origin," 223–260; Arvind Rajagopal, *Politics After Television: Religious Nationalism and the Reshaping of the Indian Public* (Cambridge, U.K.: Cambridge

University Press, 2001); and Rajagopal, "Hindu Nationalism in the US," 467–496.

65. Vinay Lal's pioneering work in this area highlights some key facets of this history as it materializes online in the present moment. See Vinay Lal, "Aryavarta and Silicon Valley" (in his *The History of History*) and "The Politics of History on the Internet."
66. See, for example, Bosah Ebo, "Internet or Outernet?" in *Cyberghetto or Cybertopia? Race, Class, and Gender on the Internet*, ed. Bosah Ebo (Westport, CT: Praeger, 1998), 1–12; Mark Poster, "Cyberdemocracy: The Internet and the Public Sphere," in *Virtual Politics: Identity and Community in Cyberspace*, ed. David Homes (London: Sage, 1997), 212–28; Howard Rheingold, *The Virtual Community: Homesteading on the Electronic Frontier*, rev. ed. (Cambridge, MA: MIT Press, 2000); and Ziauddin Sardar, "alt.civilizations.faq: Cyberspace as the Darker Side of the West," in *Cyberfutures: Culture and Politics on the Information Superhighway*, ed. Ziauddin Sardar and Jerome R. Ravetz (London: Pluto Press, 1996), 14–41.

CHAPTER 2

THEORIZING SOCIETY: FIELD, CLASS, CAPITAL, AND THE STATE

Bourdieu's theoretical project is characterized by a radical skepticism about the notion of the autonomous human as the subject and object of knowledge and about the possibility of unfettered epistemological access to a realm of ontological authenticity. But Bourdieu does not uncritically accept the claim that structuralist theories, which emphasize the epistemological primacy of objective structure over subjective consciousness or action, can sufficiently explain social phenomena. Rather, Bourdieu seeks to overcome dualistic conceptions of structure and agency, subjectivism, and objectivism.[1] Brubaker views Bourdieu's project as ultimately rooted in a metatheorization of the relationship between "mind-independent" and "agent-independent" factors and conditions (i.e., objective conditions) and the subjective realm of "individual beliefs, dispositions, and practices."[2] Bourdieu's metatheoretical framework brokers the claims of (a) external conditions and factors that shape agents' lives and (b) the undeniable logic of subjective experience, the "objectivity of the subjective" and "the reality of representation," to reveal the dynamics of how these dimensions relate to each other in constituting social experience. This metatheoretical objective permeates the theorization of all his key concepts.[3]

Bourdieu's theoretical framework may be termed descriptive in that it does not ultimately seek causal motivations in its explanation of social phenomena. The claim underlying Bourdieu's elucidation of social phenomena through theoretical categories is not that a person, group, or society *actually* functions in this or that manner. The claim, rather, is that a person, group, or society can be viewed *as if* they were functioning in accordance with certain theoretical concepts. In his theorization of classes, for example, Bourdieu emphasizes that he is speaking of a "class on paper" that is solely an explanatory device and "not really a class, an actual class, in the sense of being a group, a group mobilized for struggle."[4] There is no attempt to conflate the theoretical framework with the object of research.

For Bourdieu, it is also essential that any framework must foreground its own status as theory. Concurrently, his antifoundational stance does not preclude the possibility of "objective" and scientific analysis. Objectivity and scientificity in Bourdieu's schema are important objectives that should guide the research endeavor. An objective scientific analysis rests on the following methodological principles: a reflexive examination of the assumptions underlying the researcher's worldview and a stringent demystification of the assumptions underlying the worldview of those who inhabit the social universe being studied, adequately historicizing the object of inquiry, and translating historical inquiry into valid sociological generalizations.

The framework employed in this book, which draws on the work of Bourdieu, can be separated into a qualitative component, comprising concepts that enable a historical reading of social phenomena, and a quantitative component that empirically tests manifestations of those phenomena. This chapter details the qualitative component, including (but not limited to) four key concepts from Bourdieu's work—class, field, capital, and the state—that provide the theoretical lens for reading the historical relationship between technology and nationalism in India. Following an explication of the key concepts in Bourdieu's work, I present a brief literature survey of scholarship based on Bourdieu's work. The chapter also undertakes an engagement with three critically relevant issues for my analysis: the problematic of instrumentality in Bourdieu's theory; the problematic of Eurocentricity; and the problematic of the religious field.

Key Concepts in Bourdieu's Work

Habitus, Practice, Dispositions, and Doxa

Bourdieu uses the term *doxa* as early as *Outline of a Theory of Practice* in relation to his theorization of the *habitus*.[5] The concept of the habitus may be understood as an explanation of the functioning of social space at the micro level, a description of the relationship between "a particular type of environment" shared by a group of people and the practices of those who inhabit that shared space.[6] There are structures that shape the character of particular shared environments. For example, in class societies, it is "material conditions of existence" that constitute the respective social spaces inhabited by different classes.[7]

Practice, an important term in Bourdieu's work, can be defined as those embodied activities and competencies that are "learned" and carried out by individuals in a social space. But this learning is not of the order of something that is consciously incorporated by an individual into his or her repertoire of responses, actions, or reactions; neither, for that matter, does this learning operate as an unconscious motivational basis for all practices. It would be more appropriate to say that these practices are acquired as a result of being integrated, acclimatized, and shaped in a particular type of environment. These learned practices, in turn, enable individuals to negotiate interactions with other individuals in that social space.

Bourdieu argues that the structures that typify social spaces give rise to *dispositions* in the members of a social space. Dispositions can be understood as inclinations toward certain responses, as the tendencies to make one choice over another and to privilege one action over another, that is, the tendencies to regularly engage in certain practices as compared with others. The habitus is a system of such dispositions that endure across space and time. An individual may inhabit more than one habitus, and various habituses may overlap to some extent. However, any particular habitus is circumscribed by a group's homogeneity. Operating as a worldview—a framework of cognitive apprehension, moral judgment, ethical commitments, or aesthetic inclinations—the habitus becomes the basis for enacting that worldview through practices.

However, what makes a particular habitus distinct from another habitus, and what makes a habitus an "objective" basis for engendering certain dispositions and practices in all those who inhabit that habitus, is the fact that there is a range of practices and dispositions for any particular habitus, which corresponds to what is thinkable within that habitus. There is a limit to the possibilities "allowed" by the perceptual framework corresponding to any habitus. What sets this limit and lies beyond it is what Bourdieu terms *doxa*. To question the doxa is an act essentially in the order of heresy, for it is to question the very basis on which not just particular practices or dispositions ultimately rest, but on which the very system that is the basis of *all* practices in a habitus ultimately rests. Hence, Bourdieu argues, for those who inhabit any particular habitus, what counts as liberal, radical, conservative, or orthodox is *all* within the realm of the thinkable, that is, within the ambit of what does not challenge the doxa. The doxa may be viewed as akin to a substratum of presuppositions, and the acquired practices and dispositions within a habitus may be seen as reflections, albeit unselfconscious, unarticulated, or untheorized, of taken-for-granted deductions about reality itself.

Field and Capital

Bourdieu employs the notion of the *field* to explain the mode of functioning of, and the composition of, social space at what may be called the macro level in a society, in contrast to the habitus, which theorizes practices at the micro level. Social space can be understood as being made up of different and distinct (although often overlapping) fields, which correspond to different spheres of activity and practice, such as the cultural, economic, social, and political. However, this conceptualization of social space is not just a theoretical grid that explains social space in the present context; the existence of distinct domains of social life has a real historical basis in the mode of development of Western societies "since the middle ages."[8]

A particular field in a society can be viewed as an embodiment of the valuation of, exchange of, and struggle over the resources of the field between different groups of inhabitants in the society.[9] The resources that are deemed valuable in any particular field may be understood as a valid kind of currency in a particular field. It is these field-specific resources that Bourdieu defines as *capital*. The power relations that structure a field do not necessarily operate as an unequal distribution of resources, for each group of inhabitants can surely bring its "own" resources into the field. Rather, the relations that structure a field operate through the legislation of what kinds of resources count as a valid currency of exchange, that is, what kinds of resources translate as valid capital for the field. For example, in the cultural field, while each group may bring its own set of cultural practices into the field and each group may possess equal resources in this basic sense, how much capital each group possesses is decided by what qualifies as culture within that group's repertoire of resources. The criteria for what counts as culture is decreed by the dominant class in that field, which is the class that possesses the most cultural capital and whose interest that particular structure of the field serves. So any group seeking to improve its relative standing in social space, by aspiring to a position of greater power than it has had before, reinforces the dominant definition of culture and the very structure that serves the dominant group's interests.

What are negotiated and contested in the exchange of capital within a field are not just those actions that would allow various groups to increase their capital, but equally importantly, the very stakes by which capital is defined at all, which Bourdieu terms *nomos*. Nomos is defined by Bourdieu as the irreducible, foundational, "fundamental law" that structures a field.[10] If the struggle for accumulation of capital within a field can be understood as a game played by participants according to the particular rules of that field, then the nomos of the field is the foundational principle that embodies and determines the field-specific rules and stakes. Each field has its own nomos, which reflects the relatively autonomous logic of the field. The nomos may also be understood as the regulative principle that orders the functioning of a field. Since it is the constitutive structure of a field, it is not dependent on any forces within the field. It does not have to be explained in terms of internal forces, nor can it be questioned from within the ambit of the field. The nomos is essentially invisible to those in the game.

Yet the nomos is neither a transcendental eternal idea nor a principle of abstract logic. It is a historically shaped view that reflects the interests of the groups that hold dominant positions in a field. It is, in this sense, arbitrary because there is no necessary or intrinsic reason for one principle as opposed to another to orient the functioning of the field. The nomos demarcates the limit to what is thinkable within the field even as the nomos must itself remain outside the ken of the thinkable.[11] Indeed, to identify the nomos is to bring to light the fact that the nomos serves particular interests. Consequently, it is in the interests of the most privileged groups of society to continually invest in the game to legitimize the principle of accumulation of capital that benefits them.

In any field, what different groups challenge each other for is not just an increase in the amount of capital they possess, but also the criteria by which something is considered genuine capital and for the corresponding right to define that nomos. What this implies is that all groups who are participants in a field share the assumption that increasing their capital requires participating in the game of exchange, negotiation, and contestation that takes place in the field. Bourdieu states that to participate in the game is to tacitly accept the rules of the game. It is to be "interested" and "absolutely invested" in the game.[12] Participation, in turn, is an affirmation of the stakes according to which the game is being played. Even if a player wishes to challenge the stakes by which capital is defined in a field, he or she can only do so by playing according to those very stakes. The participation in this game even by those classes against whom the dice is loaded amounts to a reaffirmation of the structure of the field, that is, the reaffirmation of those practices that serve a dominant class as the most authentic incarnation of that sphere of social activity. In other words, this amounts to a "recognition" of the nomos of the field.

Bourdieu also argues that value or capital in one field does not necessarily translate into value in another field. Nor, for that matter, can capital in any field be understood in terms of its economic value, that is, in terms of economic capital. This is a function of two related facts about the nature of fields: first, that fields are relatively autonomous from each other, and, second, that the nomos—the "law" that structures each field and dictates the principle for the struggle over capital in that field—cannot be understood or represented in terms of any simple equation. There is no fixed formula that can explain why one kind of capital translates (or fails to translate) into another kind of capital; there is no fixed relationship between the various types of capital. The nomos of each field is arbitrary and can be understood appropriately only in terms of the history of a particular field in a particular society.

Nevertheless, certain kinds of capital are "convertible" into other kinds of capital; for example, economic capital can be converted into educational capital and educational capital into social capital. While there is no general equivalency between forms of capital, there are rates of exchange according to which particular form of capital may be converted to another particular form of capital. The mode and mechanisms by which, say, economic capital translates into social capital and the factors that determine this translation will vary from society to society. What emerges from this scenario is that in a particular society some kinds of capital may be more advantageous than others in terms of their ability to convert into other kinds of capital. This creates a third form of "struggle" over capital: in addition to (a) the volume of a given capital and (b) the definition of what will be valued in a given field (nomos), there is (c) an attempt to control the relative advantage of forms of capital in relation to one another, that is, the ease of transfer of capital from one field to another.

Types of Fields and Capital

Bourdieu's examination of types of fields and capital is distributed over detailed book-length empirical studies of particular fields in the French context and a number of essays that offer theoretical illuminations on various aspects of fields and capital.[13] Broadly, one can define the following distinct categories of fields and capitals. Bourdieu argues that "in a social universe like French society, and no doubt in American society," the "fundamental social powers" are, in order, economic capital, cultural (or informational) capital, and social and symbolic capital.[14] By the economic field, Bourdieu is referring to the domain in which capital, in the classic sense of the term, is produced, exchanged, and accumulated. The economic field refers to the "market economy based in capitalist principles" that has been "separated out and constituted as a relatively distinct sphere of production and exchange" in Western European societies.[15] Cultural or, more broadly, informational capital "is typically marked by socially but differentially recognized and constructed qualifications, not least those formally given by education but also other forms of social attainment to which rank might be attached."[16] For Bourdieu, social capital and symbolic capital are "strongly correlated," the former consisting of "resources based on connections and group membership" and the latter of "the form the different types of capital take once they are perceived and recognized as legitimate."[17] Symbolic capital, as defined by Fine, refers to "prestige, as in honour."[18] The political field pertains to "the narrow sense of 'politics', i.e., the sphere of political parties, electoral politics and institutionalized political power."[19] While the educational field overlaps with the cultural field (or both might be subsumed under a broad "informational" field), it can be conceptualized more narrowly as a distinct field concerned with the transmission of knowledge and credentials through institutional mechanisms such as school and university systems. One can also conceptualize fields of intellectual and artistic production that, in the course of Western European history, "emerged and acquired a certain autonomy, with their own institutions (universities, museums, publishing houses, etc.), their own professionals (intellectuals, artists, writers, etc.) and their own principles of production, evaluation and exchange."[20] Scientific capital is described by Bourdieu as the capital associated with specialized scientific skills and noteworthy scientific achievement.[21] Scientific capital does not refer to the cultural, economic, intellectual, or symbolic capital that a scientist would possess by the very act and fact of being a scientist. Rather, scientific capital is a surplus, specialized form of capital gained from recognition from one's peers in the scientific field for a special achievement over and above regular professional obligations and output. It is an affirmation for having "distinguished oneself positively by a *distinctive contribution*."[22] This "plus" or surplus capital is described by Bourdieu as "incorporated cultural capital."[23] The fact that scientific capital is contingent upon the recognition of other scientists marks it, strictly speaking, as different from symbolic capital (which may possibly be accrued in the scientific field without such peer recognition), although the two "are closely linked."[24] "One of the key writings in which Bourdieu developed the notion of the field" is his essay, "Genesis and Structure of the Religious Field."[25] Bourdieu's theorization of the religious field draws on the work of Durkheim, Weber, and Mauss, while also pointing to the limitations of the work of these theorists.[26] The religious field refers to the domain of organized religion, wherein religious authorities control the nomos or stakes of the field.

Class

I have until now used the generic term *groups* to describe the different actors engaged in the struggle over capital in fields. Bourdieu defines *classes* as groups of people that share a position in the hierarchy of capital in a field. Classes may also be defined in terms of the overall amount of capital possessed across fields. However, this does not necessarily mean that a group of people who share a position in one field will share the same position in another field. A person may be located in a dominant position in the cultural field based on the amount of cultural capital he or she possesses, while he or she may occupy subordinate positions in other fields, say, the economic or the educational. On the basis of Bourdieu's theorization of class, two persons can be located in the same segment in one field—say, the cultural—while they may hold different positions in the hierarchy of other fields. Social space is not composed, as in the Marxist vision, of discrete classes that are separate or opposed to each other. Instead, social space is composed of a network of *possible* contestations and alliances between groups across different fields.

This implies that no two classes, regardless of the difference in their quanta of capital, will be necessarily opposed to each other. And no two classes, no matter how close, will be necessarily allied with each other. On the other hand, it is never theoretically impossible that even the two "most separated classes" will join together in an alliance.[27] It is important to draw attention here to Bourdieu's notion of class as a theoretical concept. Bourdieu uses the concept to describe classes "on paper" that cannot be enumerated as real classes with a perfect one-to-one correspondence with empirical groups that categorically define themselves in terms of class position.[28] At most, his notion of a class may be thought of as a "probable" class "in so far as it is a set of agents which will place fewer objective obstacles in the way of efforts of mobilization than any other set of agents."[29]

Bourdieu's Theory of the State

Bourdieu understands the state as the "*culmination of a process of concentration of different types of capital*: capital of physical force or instruments of coercion (army, police), economic capital, cultural capital or (better) informational capital, and symbolic capital)."[30] As fields emerge historically, shaped by the internal play of capital, the state's accumulation of the capital pertaining to each field increases as well. With the state becoming the possessor of significant amounts of different kinds of capital, it becomes the holder of a kind of "metacapital," which guarantees authority over each of the fields, particular species of capital, holders of capital in different fields, and exchange rates between species of capital.[31] It follows that the stakes for playing the game within or across fields, and inscribed within the nomos of each field, will be, in part, structured by the state.

The exercise of metacapital, or statist power, is effected, partly at least, through what may be termed the "objective" aspects of the state—institutions that enforce and affirm the presence of the state, such as organs of the state (e.g., the judiciary), policy-making bodies (e.g., a federal or central government bank), or administrative bodies (e.g., a central tax department or revenue service). But these "objective" aspects of the state do not merely exist over and above the citizens of the state as constraints upon the actions and perceptions of citizens. Bourdieu points out that "the state exists in two forms: in objective reality, in a form of a set of institutions such as rules, agencies, offices etc., and also in people's minds."[32] The heart of Bourdieu's argument is as simple as it is profound. Any particular state can be described as the outcome of a shared history between state (or structures of government) and people; a shared history of state and people (sometimes perceived as state *against* people) can, in turn, be understood in terms of struggle over capital in different fields. Bourdieu's proposition that the state exists in "subjective" form suggests, via the notions of inculcation in the habitus and position-taking in the field, that the categories and structures of cognition and perception of the citizens of a state are historically constituted through a shared relationship between state and people. As Bourdieu puts it,

> The construction of the state is accompanied by the construction of a sort of common historical transcendental immanent to all its subjects. Through the framing it imposes upon practice, the state establishes and inculcates common forms and categories of perception and appreciation, social frameworks of perception or of understanding of memory, in short *state forms of classification*. It thereby creates the conditions for a kind of immediate orchestration of habitus which is the foundation of consensus over this set of shared evidences constitutive of (national) common sense.[33]

This understanding of the subjective form of the state has several practical implications. The first implication is that the people governed by a state are disposed, in the manner of those disposed within a habitus, to respond to the state, which is in reality the position-taking strategies of all of the people objectified. Players within each field and across fields, engaged in the struggle over different types of capital, "recognize" the terms of each game but also "recognize" the fact that they are playing *all* games on terms that have been orchestrated by the state. In this recognition, they contribute to the affirmation of the nomos proposed by the state within and across fields. Just as the habitus is embodied within the inhabitants of that habitus in the form of dispositions, so is the state incorporated in its citizens. The state, in this manner, shapes structures of perception and cognition across the society that the state governs. This is what Bourdieu means by the phrase "Minds of State," suggesting that the state exists as much as an entity "outside" of its citizens as it exists "of" the citizens.[34] It is this incorporation within the state, which is shared by all the citizens of the state, and the incorporation of the state within the citizens that would explain why citizens would be disposed to comply with the state's dictates.

The "common historical transcendental" that Bourdieu speaks of may be viewed as a common language that the state and people speak. While the particularity of such a common transcendental depends on the particular history of a state, and while the state defines the character of this transcendental to a greater extent than the people, the common transcendental refers, even if obliquely, to the shared history of a particular state and people. In other words, it is the particularity of the relationship that is invoked in the struggle between the state and the people for the accumulation of capital and for redefining the nomos of any field. For example, in one country, the struggle for accumulation of political capital (the struggle to make certain issues part of the agenda of governance) may historically have been founded on the platform of human rights. In other countries, the struggle for accumulation of political capital may invoke constitutional authority, whether or not that authority is ultimately grounded in the notion of human rights.

The Material and the Symbolic

In theorizing any and all of his concepts, Bourdieu does not distinguish between the material and the symbolic. Whether one is analyzing practices, capital of different kinds, or the nomos, the materiality and symbolic significance of the referents of these terms cannot be separated from one another. This is entirely consistent with Bourdieu's metatheoretical objective of overcoming the subjective-objective dualism characteristic of structuralist and subjectivist social theories, which would impute distinct ontologies to the domains of the material and symbolic. For Bourdieu, the material and symbolic are mutually constitutive, yet not in a dialectical manner as distinct elements structured in a symbiotic relationship. Rather—if one were to analytically distinguish between the two for the purpose of argument—the material and symbolic suffuse each other in the very mode in which they are traded for capital, acted upon in the realm of the habitus and practices, internalized and expressed as dispositions, configured as the doxa or nomos, and so on. Inasmuch as we can speak of the two realms as separate, Bourdieu argues that the complex co-constitutive relationship between the two means that we can see in them an ontological synchronicity or harmonization.

In attempting to transcend the pitfall of ontologizing the material and the symbolic, Bourdieu's framework seeks to avoid the minefield of "authenticity," that is, the question of where the "real" meaning or significance of any event lies—in the thought or the action, in the behavior or its interpretation by an individual or group. Bourdieu's framework does not deny the possibility of the "authentic" meaning or significance of an object or practice for any individual or group. Indeed, individuals and groups can be viewed as vested with the agency to construct meaning for themselves. What the framework seeks to dismantle or bypass is the burden of the theory necessarily having to proffer a definition of what constitutes authenticity. The framework, then, does not dismiss the value of the meaning individuals and groups find and express in actions, objects, and practices. To the contrary, it recognizes the legitimacy of these individual or social claims to meaning while refusing to pass evaluative judgments on them. What the framework does is recast these questions such that the focus is on the structures and mechanisms by which "meaning" is constructed at all. In this regard, neither the material nor the symbolic possesses epistemological or ontological priority over the other.

Bourdieu speaks of symbolic capital as a kind of *uber*capital (prestige, recognition, or honor) that derives from the accumulation of a critical mass of other kinds of capital in any given social setting. This may give rise to some conceptual confusion, given that (a) Bourdieu defines symbolic capital as comprised of intangibles, such as prestige or, more generally, recognition, yet (b) Bourdieu claims that the material and symbolic do not possess ontological autonomy from one another. Symbolic capital for Bourdieu is a kind of universal currency that can be readily cashed in any field, granting its holder a relatively privileged or dominant position as he or she enters into the game of capital in any field. The term *symbolic* here is best understood as a descriptor to indicate capital that is not yoked or limited to the nomos of a particular field. Yet this symbolic capital itself has been accumulated earlier in the context of particular fields where it is not outside the dynamics of "material" exchange. The ontology of symbolic capital, therefore, reflects the same mutually constitutive pattern of the material and the symbolic. Similarly, its implementation occurs in the context of particular fields that operate in terms of both the material and symbolic. In the same vein, the benefits that may accrue from symbolic capital are also constituted of both these dimensions.

The Individual and the Social

Finally, a note on Bourdieu's notion of the "individual" should be provided. For Bourdieu, the "individual" as a concept is meaningful only in reference to the social space inhabited by that individual. Bourdieu's individual is not an individual in the sense of a "person" or "being," but is an organism that is defined by the matrix of possible practices that characterize a particular social space. Bourdieu's theory is not predicated on a priori notions of the social, either. Social classes are mappings of distributions of types of capital across social space. Bourdieu's theoretical framework does not propose an ontology of the social but describes patterns of interestedness across society that correspond to a network of social relations.

LOCATING BOURDIEU ACROSS DISCIPLINES

Bourdieu's scholarship is

> remarkable for its theoretical sophistication and for its ethnographic acuity, and [it constitutes] one of the most significant of recent attempts to adapt the theoretical legacy of classical social

> theory to the empirical study, from a broadly critical perspective, of contemporary society.[35]

However, as Pileggi and Patton note, Bourdieu has gone "mainly under- or misread in the Anglophone academy."[36] Honneth states that "Bourdieu's theoretical work is only just beginning to make an impact" in West German, British, and American social science.[37] Yet, over the course of more than two decades, Bourdieu's cumulative influence on various fields of scholarship in different locales and traditions of scholarship has not been inconsiderable. In numerous areas, including the sociology of education, anthropology, philosophy of knowledge, cultural studies, social theory, literary studies, and historical scholarship, Bourdieu's work has been theoretically and empirically examined, applied, developed, tested, and critiqued.[38] This brief survey presents engagements with Bourdieu's work in three general areas—social theory, cultural studies, and the sociology of education—where he has been especially influential. The objective of the survey is to present perspectives on Bourdieu's specific contribution to these academic fields and to outline the chief grounds on which his work has been critiqued. Based on these critiques, I will summarize the key theoretical challenges to Bourdieu's work with which my analysis needs to engage.

Bourdieu as a Social Theorist

In social theory, Bourdieu has predominantly been viewed as a theorist of class and culture, typically as either a Marxist or a Weberian.[39] Huang, in his analysis of rural class struggle in China from 1946 to 1979, for instance, credits Bourdieu for incorporating the representational or symbolic domain along with the objective domain in class analysis, yet critiques him for privileging the latter realm.[40] Huang sees Bourdieu as an important figure within the Marxist tradition, but finds Bourdieu's theory of class ultimately guilty of the objectivist Marxist fallacy. For Honneth, Bourdieu is essentially a Marxist theorist, though Bourdieu's main emphasis is on the cultural-symbolic realm, which doctrinaire Marxists might see as secondary to the processes of actual social relations.[41] Honneth's chief objection to Bourdieu's theory is that it cannot explain the substantive commitments held by different social groups or classes to their definitions of capital or, generally, to their way of life.[42] Arguing that Bourdieu's theory holds largely for the economic field, Honneth describes Bourdieu as a Marxist who applies an economistic perspective to all spheres of social life.

Joppke broadly locates Bourdieu's understanding of class in the Weberian mode of "entirely socially constituted competitive groups,"[43] noting that the family is the basic unit of class in Bourdieu, since it is from the family that capital is inherited. Class struggle is vested in the attempts of families striving to maintain (or reproduce) and improve their social position from one generation to the following. Joppke lauds Bourdieu's notion of class struggle for its originality, but, like Honneth, argues that Bourdieu sees "economic calculation" as the basis for all aspects of social life.[44] He is also critical of Bourdieu's framework for its inability to recognize and theorize the possibility of radical social change.[45] Gartman's neo-Marxist critique of Bourdieu levels a similar charge, proposing that his theory is unable to recognize human agency or explain the possibility of revolutionary activity.[46] He points out that it also fails as a Marxist theory since it "rejects the Marxist emphasis on production of goods in favor of a Weberian emphasis on positions of goods."[47] Additionally, the theory tends to overemphasize the importance of the cultural domain and does not take into account the concrete historical realities of capitalism. For Frow, too, in Bourdieu's deterministic framework, there is no escape from the all-pervasive grasp of dominant structures of practice or perception, as a result of which there is no theoretical scope for "critique and social transformation."[48] Frow also contends that Bourdieu does not satisfactorily explain his own position as observer and researcher of the social world and that his work implicitly claims to be written from a position above and beyond society.

Grusky and Sorensen see Bourdieu's theory as an attempt to solve a particular problem in Weberian class analysis—that of the unification of class and status.[49] They commend Bourdieu's ambitious attempt to resolve this tension, but ultimately find it unsatisfactory. Their main objection is that Bourdieu's empirical work emphasizes the cultural rather than economic "implications of occupational desire."[50] In a comparative assessment of

Bourdieu, Weber, and Karl Mannheim with regard to their perspectives on the sociology of knowledge, Ringer speaks to the value of Bourdieu's range of concepts such as habitus, doxa, and capital and his differentiation of species of capital for the analysis of the intellectual field.[51] However, Ringer points out that Weber's concept of status as distinct from class may be more useful than Bourdieu's notion of cultural capital in understanding what kinds of norms and practices are legitimated in fields and systems as dominant, since the Weberian perspective is more resolutely historical than Bourdieu's model of social stratification.[52]

For Brubaker, Bourdieu, judged by his theory of class, is as distant from Weber as from Marx.[53] Class and habitus are the two foundational theoretical concepts in Bourdieu's work, the former operating as a "universal explanatory principle," which is what distinguishes him from both Marx and Weber, since for them the concept of class served "definite and limited explanatory aims."[54] Brubaker argues that Bourdieu's social theory is characterized by a "strong" tension between a theoretical "impulse towards generality" and the "concrete richness of accounts of particular practices, institutions, and rituals" found in Bourdieu's empirical studies.[55]

Bourdieu in Cultural Studies

Bourdieu's reception in cultural studies has been more varied than his reception in social theory, possibly as a result of the nature of the field. In his work, Bourdieu maintains a strategic distance from a categorical affiliation with cultural studies.[56] Yet Bourdieu's emphasis on reflexivity and his demystification of the conditions of production of practice also point to an affinity with cultural studies.[57] While in the early phases of cultural studies, Bourdieu has often been viewed as a Marxist, more recently his theoretical framework appears increasingly to have been employed on its own terms in the area.[58]

Pileggi and Patton note that Bourdieu's work embodies the apparently paradoxical imperative of "both *autonomy* and *engagement*."[59] The academic field requires autonomy as the basic condition of possibility for the production of academic knowledge, but autonomy without engagement leads to ossification.[60] Intellectuals must thus situate their position and work in relation to the larger world in order to resist and reveal the doxa of the academy. It is this model of intellectual work, exemplified by Bourdieu, that is especially valuable for cultural studies.

Calhoun describes Bourdieu as "one of the two most important sociologists of culture," the other being Raymond Williams.[61] While Bourdieu's work does not deal with critical issues in the field of cultural studies, he can be viewed as a sociologist who has made " 'culture' central to understanding the whole of social life."[62] For Le Hir, Bourdieu's "reflexive sociology, like cultural studies, falls between disciplinary cracks."[63] As with cultural studies, Bourdieu's emphasis on reflexivity allows for flexible and "strategic" modes of political intervention and resistance.[64] While there are important differences between cultural studies and reflexive sociology, such as the former's "happy" equation with poststructuralism as contrasted with the latter's antipathy to it, these do not necessarily indicate the fundamental incompatibility of the modes of academic inquiry.[65]

Garnham and Williams view Bourdieu's work as an important intervention in a cultural studies debate between two Marxist traditions—the "culturalist" Marxism represented by the early work of the Birmingham Centre for Contemporary Cultural Studies and the "theoreticist" Althusserian and Lacanian Marxism, which emphasizes the analysis of ideology in the "text"—and argue that Bourdieu's work dialectically transcends these positions.[66] They see Bourdieu as unflinchingly committed to a materialist perspective on struggles, including the significance of symbolic struggles. But they find a "functionalist/deterministic residue" in his notion of reproduction.[67] They also problematize Bourdieu's own work, questioning whether it evades or reproduces the hierarchies of the intellectual field. Bourdieu's notion of popular culture has also been critiqued by Rigby for its failure to recognize the democratizing impulse of the popular culture movement.[68] Rigby argues that Bourdieu's assessment of working class culture fails to recognize its autonomy and revolutionary potential.

Bourdieu in the Sociology of Education

In the sociology of education, Bourdieu is often viewed as a reproduction theorist offering a testable social scientific model of how class differences are perpetuated by the educational system from generation to generation. This is not surprising considering that "class inequality in schooling is the most enduring topic in this subdiscipline."[69] Most of the readings of Bourdieu in the sociology of education aim to test the validity of his model through empirical testing and quantitative analysis.

Robinson and Garnier view Bourdieu's work on the French education system as a reproduction theory of inequality.[70] Their chief objection to the theory is that it overemphasizes the role of class background (parental influence) in reproducing class privilege and does not sufficiently take into account the gendered nature of this process. Persell and Cookson qualify Bourdieu's analysis in suggesting that the mode in which cultural capital is appropriated may be as important as the appropriation of cultural capital per se in perpetuating class reproduction.[71] Lamont and Lareau clarify Bourdieu and Passeron's model of the relationship between cultural capital and social reproduction and propose a new definition of cultural capital.[72] They point out that across the range of Bourdieu's work, cultural capital takes on several different meanings; it is, variously, a "power resource," synonymous with "taste," and an indicator of "class position."[73] They advocate a more flexible notion of cultural capital and suggest that cultural reproduction should be viewed as just one instrument of social exclusion. They also question the applicability of the model to societies other than French society. Archer, similarly, critiques Bourdieu for refusing to acknowledge the particularities of the French system and for presenting the French case as universally valid.[74]

According to Erikson, Bourdieu's theory is "a life course analysis of class reproduction" that is at once "a model of class structure."[75] Erikson posits that Bourdieu ignores two crucial dimensions of class structure: social networks and "class relations at work."[76] Erikson argues that Bourdieu's theory ignores the specificity of individual life experience which may vary significantly across generations and that his concept of class also neglects dimensions of gender, race, and ethnicity. Other critiques of Bourdieu as a reproduction theorist stress that parental occupation and cultural capital (as Bourdieu defines it) are not sufficient for explaining the reproduction of class privilege.[77]

In sum, assessments of Bourdieu as a reproduction theorist argue that in his work class is granted theoretical primacy over other variables (most notably gender, but also race and ethnicity). His definition of class is criticized as overtly deterministic in its emphasis on cultural capital as a foundational category. Scholars also argue that Bourdieu's definition of culture itself needs to be made more expansive and flexible.

KEY THEORETICAL ISSUES

This brief survey outlines several theoretical challenges to Bourdieu's framework from different areas of academic inquiry. There are three issues that are of particular significance for the purpose of this book. Two grounds on which Bourdieu's theoretical framework has most commonly been critiqued are (a) its instrumentality or utilitarian bias and (b) its generality and normative assumptions, which can be reframed as the problematic of Eurocentricity. Additionally, there is a third ground—that of the theoretical adequacy of Bourdieu's conceptualization of the religious field. Here, too, Bourdieu is susceptible to charges of instrumentality and Eurocentricity. My objective in this concluding part of the chapter is to outline a general approach to applying Bourdieu's concepts in a manner that may answer critiques of his work on each of these grounds.

The Problematic of Instrumentality

Critics of Bourdieu state that his perspective is fundamentally instrumentalist or utilitarian. This reading pits a domain of instrumentality against a corresponding noninstrumental realm, the latter consisting of practices that possess meaning in and of themselves, which are pursued without interest outside the calculus of capital. The charge against Bourdieu is that he refuses to recognize this realm, seeking utilitarian motive in all social action, belief, thought, behavior, and practice.

In *Practical Reason*, Bourdieu takes up the question of whether a "disinterested" act, outside the play of capital and free from the conscious desire for reward, is possible at all.[78] He argues that "One can be interested in a game (in the sense of not indifferent) while at the same time being disinterested."[79] This seemingly paradoxical statement becomes clear when Bourdieu's notion of "interestedness" is explained. Interestedness is not conscious action toward a well-defined end. Rather, interestedness is a sense—which has been embodied physically and mentally within all social actors—that the game is of value. As Bourdieu points out, "That is what I meant in speaking of interest: games which matter to you are important and interesting because they have been imposed and introduced in your mind, in your body, in a form called the feel of the game."[80] For the game to be effective, players, in fact, are not consciously in the know that they are playing games: "Social games are games that are forgotten *qua* games."[81] Every field requires participants to be interested and invested in the game. Entering the field is a recognition of the stakes and a willingness to play the game. Participants may have a diverse range of "motives" or "views" for entering the field. Yet, the very fact that they enter into the field is a recognition "that it is worth the effort to struggle for the things that are in play in the field."[82]

In this sense, disinterestedness as lack of a categorical objective does not mean one is not interested in the game. Those who are outside the game and do not want to play it are "indifferent," which is akin to being apathetic.[83] To view interestedness as instrumentality or utilitarian action may be to misread it as a kind of rational choice action. It may be more adequate to say that while Bourdieu's framework lends itself to being interpreted as an instrumentalist perspective, it is by no means exhausted by that perspective.

The Problematic of Generality or Eurocentricity

Bourdieu has also been critiqued for universalizing the particularities of French society into a normative framework for the analysis of society at large. At the same time, Bourdieu also attempts to be fair to the contextual specificity of any social setting in his empirical analyses. This tension manifests itself as two theoretical problems: first, in Bourdieu's own work, as Brubaker has noted, there is a disjunct between the "extreme generality" of his theory and the richness of his particular empirical studies; second, with regard to applications of Bourdieu's work by other scholars, the theory itself appears as inadequate and deficient for assessing the social setting in question.[84]

Given the focus of this book, the problematic of generality or normativity can be broadly recast as one of Eurocentricity. Some of the critiques explicated earlier imply that Bourdieu's theory does not necessarily travel well from one society to another. Such a shortcoming would appear to be compounded in the case of the theory having to travel to a non-Western postcolonial context, given that concepts such as field, capital, and state are based on Bourdieu's descriptive reading of the history of Western societies since the time of the Middle Ages. An example will clarify this problem of applicability. Bourdieu argues that the state is the result of a history of shared struggle between state (or governmental structures) and people. Under colonial government, however, the very possibility of the colonized population engaging in a shared struggle with the state is severely limited. The colonial British administration, for instance, denied Indians entry into the political field on equal terms for control over Indian territory. The possibility of accumulating capital in the political field or altering its nomos was contingent more on one's race—that is, whether one was European or "native" Indian—than on the possession of other kinds of capital. An analysis of the colonial Indian state based on Bourdieu's concepts would need to address this fact of colonial difference, while not reducing it to a subcategory of class. Second, the struggle of Indians, with the emergence and growth of anticolonial nationalism, was for complete control of the state—to do away with the colonial state and replace it with an "authentic" Indian state. Conversely, with respect to the realm of culture, Chatterjee has argued that Indians sought to deny the colonial state the possibility of impacting or shaping the nomos of the cultural field.[85] According to Chatterjee, anticolonial nationalism proclaimed its sovereignty in the sphere of culture and values, while acknowledging that it had been subjugated in the sphere of statecraft and sovereignty. It was here, within the domain of culture, that anticolonial nationalism sought to articulate a definition of Indianness, which anticolonial nationalism would later utilize for the purpose of mobilizing political capital. One may interpret Chatterjee as suggesting that the political and cultural fields in the period of colonialism and anticolonial struggle cannot really be separated from one another. As he points out, the difficulty in adequately grasping the character of anticolonial nationalism "arises because we have all taken the claims of nationalism to be a *political* movement much too seriously."[86]

As a partial defense of Bourdieu against the general charge of Eurocentricity, one might note that his theoretical framework for reflexive sociology is, to a significant extent, shaped as a critique of Eurocentricity. Based on anthropological fieldwork among the Kabylia community in Algeria, Bourdieu arrived at the understanding that "Western European anthropology was an instrument of 'symbolic violence.'"[87] Assumptions about the inherent epistemological superiority of the Western anthropological framework meant that the discourse "could only generate a distorted representation" of traditional, indigenous (more broadly, non-Western) societies.[88] This realization led to a "shift in his self-presentation from being an 'anthropologist' to being a 'sociologist.'"[89] Reflexivity is key in Bourdieu's vision of sociological endeavor. The first task for the sociologist is to uncover the assumptions underlying the researcher's perspective—the doxa of the theoretical framework. One of Bourdieu's basic theoretical objectives is to reverse the mode in which Western knowledge has typically encountered its Other—the non-West.

Given that Bourdieu proposes, as a vital methodological principle, the requirement of historicizing and provincializing the concepts employed in research, there is no reason why Bourdieu's own theoretical framework should not be subjected to the same scrutiny. Yet, while this may be necessary, it may not be sufficient for answering the objection of Eurocentricity, for there exists the possibility that concepts may be found to be lacking for analyzing the object in question. How then might one proceed with the application of these concepts? One useful way to do so is to seek recourse to the "extreme generality" and "plastic" sociology, noted earlier, of Bourdieu's framework, which may be an asset as much as a liability. What this flexible framework offers is the possibility of incorporating elements of other theoretical frameworks, as long as these are not incompatible with each other or with Bourdieu's theoretical frame at the level of fundamental assumptions. I am suggesting that it may be more fruitful to view Bourdieu's work as a metatheory that can successfully work with a range of theoretical perspectives and modes of analysis rather than as a paradigm burdened with having to explain and describe all that it encounters. For example, Bourdieu's notions of class and capital clearly stand in opposition to a classic Marxist emphasis on the predication of cultural capital on structures of the production of economic capital. However, culturalist Marxist perspectives that challenge this conceptualization and emphasize the relative autonomy of cultural production, reception, and exchange might well be usable within a larger framework based on Bourdieu's work. Similarly, the Gramscian concept of an "organic crisis," as applied to the Indian postcolonial context, can fit well within Bourdieu's theory of the state.[90] Gramsci uses the concept to describe the condition of a society where competing group interests are only barely held together by an overarching ideology.[91] The state's ideological justification for its actions are not universally legitimized or internalized by all inhabiting groups in society, giving rise to an endemic crisis where there is no established consensus or set of rules according to which different groups can compete with each other or with the state.[92] It is possible to imagine the functioning of a political field, in Bourdieu's sense of the term, that incorporates theorizations of an organic crisis in postcolonial India. The struggle of the people with the state is configured in a manner that does not allow for that struggle to take shape as a shared history. Rather, there are several distinct games being played in the political field at any point in time, each of which seeks to preclude particular groups of inhabitants (for example, "untouchable" castes, religious minorities, ethnic groups, etc.) from participation. Bourdieu's concepts of the field and capital and the metaphor of the game can, in this manner, be applied as "metaconcepts," which can accommodate theoretical tools from other frameworks.

The Problematic of Religion

Bourdieu's theorization of religion focuses on the sphere of organized religion and authority. Describing Bourdieu's theorization of the religious field, Calhoun observes that "religion, thus, only appears as such after a division of social labor sufficient to produce priests with an interest in defining the field of religious goods over which they claim a monopoly of authority."[93] Bourdieu's essay can be located in the tradition of the "classical sociology of religion" that reflects the influence as well as limits of the work of figures such as Durkheim, Mauss, and Weber.[94] As with other kinds of practices and fields, Bourdieu emphasizes the relative autonomy of religious practice and belief, even as he locates the domain of religious practice and belief within the larger economy of the operations of social power and the establishment and legitimation of social hierarchies. Bourdieu's conceptualization of religion is, accordingly, vulnerable to the same broad critique leveled against his theoretical framework—the fact that it grants theoretical priority to the instrumental function of religion in securing, authorizing, and marking power. Verter states, for instance, that "Bourdieu's vision of religion is unidimensional…Bourdieu perceives religion almost exclusively in organizational terms."[95] Engler notes that Bourdieu views religion as "anachronistic," a feature of premodern but not modern societies.[96] He also draws attention to Bourdieu's argument that in "modernity" the state has taken over the traditional role of religion. This function, which Bourdieu terms "consecration," involves the "legitimation of existing social inequalities" by masking economic disparities.[97] According to Engler, Bourdieu grants undue importance to the role of the state in modernity and exaggerates the difference between traditional and modern systems of social relations.[98] Engler argues that both these notions operate as a denial of the significance of religion in the era of modernity and that their validity is undermined by the realities of the present. In the era of globalization, the importance of the state is diminishing. And social conditions in late modernity have engendered a greater need among humans for the consolations of consecration. Insecurity, a heightened awareness of the role of contingency in human existence, and the inability of the state to assuage these anxieties account in part for the "resurgent role" of religion in late modernity.[99]

There is, additionally, one implication of Engler's critique that is particularly important for my argument. Bourdieu's emphasis on the central role of the state and his relegation of religion as the province of the premodern society appear to be predicated on the separation of church and state, for it is only when religion has been banished from the sphere of political-bureaucratic power that the state can monopolize the function of consecration, which is now secularized. Bourdieu takes the separation of church and state as an established fact, and it is in this regard that his account of religion can justifiably be termed Eurocentric. Although Engler states that "it would be facile to argue that his conception of the state is too determined by the North American–North Europe axis, too Eurocentric, or too French," the objection, I think, is anything but facile.[100] The nature of the separation of religion and state in India and the character of Indian secularism are much-contested topics. There is, in fact, no consensus among scholars on what the separation of religion and state means in the Indian context. In light of this, Bourdieu's understanding of religion may appear somewhat inadequate for the analysis of the formation of various fields in colonial India. Yet, Bourdieu himself may provide one with a way out of this theoretical impasse.

Verter argues that "to see Bourdieu's relevance for sociologists of religion, one must—quite paradoxically—turn away from his writings on religion."[101] Verter asserts that it is Bourdieu's notion of cultural capital, which is "the most complex, most novel, and analytically the most useful" of Bourdieu's various kinds of capital, that is of great value for the analysis of religion.[102] Pointing out that "cultural capital is a matter of disposition, not just acquisition," Verter theorizes a notion of "spiritual capital" as a subset of cultural capital.[103] Spiritual capital is different from religious capital in that the former is not yoked to the sphere of organized religion. Verter states that

> spiritual dispositions may be regarded as a form of cultural capital. Personal piety may be viewed as a matter of taste—in other words, as a product of social relations—and thus a marker of status within struggles for domination in a variety of contexts. Spiritual knowledge, competencies, and preferences may be understood as valuable assets in the economy of symbolic goods.[104]

The notion of spiritual capital as a kind of cultural capital is helpful for historicizing the impact of technology in various fields in colonial India. The conceptualization can enable one to analyze the specific role of religion in the constitution of the cultural and educational field, while rescuing one from having to theorize the cultural and religious fields as either entirely separate or entirely conflated realms against the grain of historical evidence. With this metatheory as a lens, qualified by an awareness of the limits of Bourdieu's theoretical concepts, and complemented with compatible conceptual tools from other theoretical frameworks, I turn to a reading of the relationship between technology and nationalism in Indian society, from the advent of colonialism to the present era of globalized technologies and media.

ENDNOTES

1. See, for instance, Brubaker, "Rethinking Classical Theory," 87–116, and Loic J. D. Wacquant, "Towards a Reflexive Sociology: A Workshop with Pierre Bourdieu," *Sociological Theory* 7, no. 1 (Spring 1989): 26–63. Brubaker posits that the subjectivist-objectivist opposition is the central problem that Bourdieu's "metatheoretical" stance seeks to overcome (91).
2. Brubaker, "Rethinking Classical Theory," 91.
3. Ibid., 92.
4. Pierre Bourdieu, *Language and Symbolic Power*, ed. and intro. John B. Thompson, trans. Gino Raymond and Matthew Adamson (Cambridge, MA: Harvard University Press, 1991), 231.
5. Pierre Bourdieu, *Outline of a Theory of Practice*, trans. Richard Nice (Cambridge, U.K.: Cambridge University Press, 1977).
6. Ibid., 72.
7. Ibid., 72.
8. John B. Thompson, "Editor's Introduction," in Pierre Bourdieu, *Language and Symbolic Power*, ed. and intro. John B. Thompson, trans. Gino Raymond and Matthew Adamson (Cambridge, MA: Harvard University Press, 1991), 125.
9. A field is neither the result of the objective logic of forces that are wholly independent of the conscious or unconscious motivations of the inhabitants nor can the existence and characteristics of a field be explained in terms of a causal outcome of the purely subjective intentions and interests of the inhabitants. The field possesses an "objective" aspect and, at the same time, embodies the subjective interests of groups of inhabitants, even as this description does not need to articulate itself in terms of objectivity and subjectivity.
10. Pierre Bourdieu, *Pascalian Meditations*, trans. Richard Nice (Stanford, CA: Stanford University Press, 1997), 96.
11.

> Once one has accepted the viewpoint that is constitutive of a field, one can no longer take an external viewpoint on it. The *nomos*, a "thesis" which, because it is never put forward as such, cannot be contradicted, has no antithesis. As a legitimate principle of division which can be applied to all fundamental aspects of existence, the thinkable and the unthinkable, the prescribed and the proscribed, it must remain unthought. Being the matrix of all the pertinent questions, it cannot produce the questions that could call it into question. (Ibid., 97)

The nomos can be thought of as the doxa at the level of the field.

12. Ibid., 11.
13. See, for example, Bourdieu, *Distinction*, for an analysis of the cultural field; *Homo Academicus*, trans. Peter Collier (Stanford, CA: Stanford University Press, 1988), for a treatment of the intellectual-academic field; *The Rules of Art: Genesis and Structure of the Literary Field*, trans. Susan Emanuel (Stanford, CA: Stanford University Press, 1996), for an assessment of the literary field; and *The State Nobility: Elite Schools in the Field of Power*, trans. Lauretta C. Clough (Stanford, CA: Stanford University Press, 1996), for an evaluation of the educational field.
14. Pierre Bourdieu, "What Makes a Social Class? On the Theoretical and Practical Existence of Groups," *Berkeley Journal of Sociology* 32 (1987): 4, as quoted in Ben Fine, *Social Capital Versus Social Theory: Political Economy and Social Science at the Turn of the Millennium* (New York: Routledge, 2001), 56.
15. Thompson, "Editor's Introduction," 25.
16. Fine, *Social Capital Versus Social Theory*, 55.
17. Bourdieu, "What Makes a Social Class?" 4, as quoted in Fine, *Social Capital Versus Social Theory*, 56.
18. Fine, *Social Capital Versus Social Theory*, 55. Part of the challenge in analyzing the exchange of capital in fields is recognizing and theorizing capital and shifts in the nomos of the field or valuations of capital. In the economic field, what counts as capital is easily recognizable—for instance, a person's liquidity, assets, creditworthiness, and investments. Correspondingly, inequality or unequal distribution of economic capital is easy to recognize. A person who owns a palatial house obviously possesses more economic capital than a homeless person. Also, while the relative value of different types of economic capital might change over periods of time—for instance, real estate might be a better investment than mutual funds or government bonds during some cycles of the economy—it is highly unlikely that these types of economic capital will become redundant and that completely new forms of economic capital will replace them. The stocks and shares of a particular company might become worthless but stocks per se as a form of economic capital are unlikely to be discarded as modes of investment in the economic structures of contemporary societies, unless there is a radical transformation, such as a revolution in which a government declares all private property to be the property of the state and declares certain monetary instruments worthless.

    Inequity or differences in the quantum of other kinds of capital may be more difficult to discern, since the very definition of what counts as capital may be subject to greater flux. Political capital is a case in point. Political capital relates directly to what count as the appropriate issues that, according to politicians and citizens, government should address. The valuation of political capital is perhaps at its greatest flux during elections and, additionally, may be subject to events, such as war, increases in immigration, or relations between ethnic groups, that become issues of domestic political concern.

In the literary field, especially, transitions in the designation of what counts as capital might be quite sudden. There might be periods when the notion of what counts as "literary" is quite stable. Yet all it might take to generate a new form of literary capital is one work that either suggests a new field of inquiry or posits a radical new method.

19. Thompson, "Editor's Introduction," 25–26.
20. Ibid.
21. Pierre Bourdieu, *Science and Science of Reflexivity*, trans. Richard Nice (Cambridge, U.K.: Polity, 2004).
22. Ibid., 55, emphasis in original.
23. Ibid., 55.
24. Ibid., 56.
25. Craig Calhoun, "Introduction," *Comparative Social Research* 13 (1991): xi. See also Pierre Bourdieu, "Genesis and Structure of the Religious Field," *Comparative Social Research* 13 (1991): 1–44.
26. Calhoun, "Introduction," xi.
27. Bourdieu, *Language and Symbolic Power*, 232. There is an implicit theory of social change here. Spontaneous social change, sparked by the opposition between groups, is improbable, given that (a) interests of groups are identical in some fields and not in others and (b) activity by any and all classes in a field affirms the principle of capital accumulation that serves the interests of the most powerful class in that field. Yet social change, even on a large scale, is always possible if (a) it is in the interests of a large number of groups that the nomos of one or more fields should be redefined or if (b) external events, such as war or sweeping legal reforms, determine a change in the principle of accumulation of capital in one or more fields.
28. Ibid., 231.
29. Ibid., 231–232.
30. Bourdieu, *Pascalian Meditations*, 41, emphasis in original.
31. Ibid., 40–41.
32. Pierre Bourdieu, "The Myth of 'Globalization' and the Welfare State," in *Acts of Resistance: Against the Tyranny of the Market* (New York: The New Press, 1998), 33.

33. Pierre Bourdieu, *Practical Reason: On the Theory of Action*, trans. Richard Nice (Stanford, CA: Stanford University Press, 1998), 54, emphasis in original.
34. Ibid., 52.
35. Brubaker, "Rethinking Classical Theory," 87.
36. Mary S. Pileggi and Cindy Patton, "Introduction: Bourdieu and Cultural Studies," *Cultural Studies* 17, nos. 3–4 (May–July 2003): 313.
37. Axel Honneth, "The Fragmented World of Social Forms: Reflections on Pierre Bourdieu's Sociology of Culture," in *Pierre Bourdieu*, ed. Derek Robbins (London: Sage, 2000), 3:3.
38. See, for instance, Derek Robbins, ed., *Pierre Bourdieu*, 4 vols. (London: Sage, 2000). The four-volume collection presents previously published work on Bourdieu from a range of disciplines and approaches.
39. My reading of literature that draws on Bourdieu suggests that there is a paucity of scholarship that engages with Bourdieu's theory of the state. This may have to do with the fact that Bourdieu offers a systematic theorization of the state and its relationship to fields and capital fairly late in his work.
40. Philip C. C. Huang, "Rural Class Struggle in the Chinese Revolution: Representational and Objective Realities from the Land Reform to the Cultural Revolution," *Modern China* 21, no. 1 (January 1995): 107–110.
41. Honneth, "The Fragmented World of Social Forms," 4.
42. Ibid., 16.
43. Christian Joppke, "The Cultural Dimensions of Class Formation and Class Struggle: On the Social Theory of Pierre Bourdieu," in *Pierre Bourdieu*, ed. Derek Robbins (London: Sage, 2000), 3:23.
44. Ibid., 37.
45. Ibid., 41.
46. David Gartman, "Culture as Class Symbolization or Mass Reification? A Critique of Bourdieu's *Distinction*," *The American Journal of Sociology* 97, no. 2 (September 1991): 421–447.
47. Ibid., 436.
48. Frow, "Accounting for Tastes," 57.
49. David Grusky and Jesper B. Sorensen, "Can Class Analysis Be Salvaged?" *The American Journal of Sociology* 103, no. 5 (March 1998): 1187–1234.
50. Ibid., 1192.
51. Fritz Ringer, "The Intellectual Field, Intellectual History, and the Sociology of Knowledge," in *Pierre Bourdieu*, ed. Derek Robbins (London: Sage, 2000), 3:66–86.
52. Ibid., 75.

53. Brubaker, "Rethinking Classical Theory."
54. Ibid., 102.
55. Ibid., 109.
56. Craig Calhoun, "Putting the Sociologist in the Sociology of Culture: The Self-Reflexive Scholarship of Pierre Bourdieu and Raymond Williams," in *Pierre Bourdieu*, ed. Derek Robbins (London: Sage, 2000), 3:60.
57. Marie-Pierre Le Hir, "Doing Cultural Studies Bourdieu's Way," in *Pierre Bourdieu*, ed. Derek Robbins (London: Sage, 2000), 3:259–260.
58. See, for example, *Cultural Studies* 17, nos. 3–4 (May–July 2003), a special issue on Bourdieu and cultural studies.
59. Pileggi and Patton, "Introduction," 316, emphasis in original.
60. Ibid., 317.
61. Calhoun, "Putting the Sociologist in the Sociology of Culture," 59.
62. Ibid., 60.
63. Le Hir, "Doing Cultural Studies Bourdieu's Way," 261.
64. Ibid., 262.
65. Ibid., 263.
66. Nicholas Garnham and Raymond Williams, "Pierre Bourdieu and the Sociology of Culture: An Introduction," in *Pierre Bourdieu*, ed. Derek Robbins (London: Sage, 2000), 3:201.
67. Ibid., 216.
68. Brian Rigby, "Popular Culture as Barbaric Culture: The Sociology of Pierre Bourdieu," in *Pierre Bourdieu*, ed. Derek Robbins (London: Sage, 2000), 3:276.
69. Scott Davies, "Leaps of Faith: Shifting Currents in Critical Sociology of Education," *American Journal of Sociology* 100, no. 6 (May 1995): 1449.
70. Robert V. Robinson and Maurice A. Garnier, "Class Reproduction among Men and Women in France: Reproduction Theory on Its Home Ground," *American Journal of Sociology* 91, no. 2 (September 1985): 250–280. The works on the French educational system referred to in the article include Pierre Bourdieu and Jean-Claude Passeron, *Reproduction in Education, Society and Culture* (London: Sage, 1977); Pierre Bourdieu, "Cultural Reproduction and Social Reproduction," in *Power and Ideology in Education*, ed. J. Karabel and A.H. Halsey (New York: Oxford University Press, 1977), 487–511; and Pierre Bourdieu and Luc Boltanski, "Changes in Social Structure and Changes in the Demand for Education," in *Contemporary Europe: Structural Change and Cultural Patterns*, ed. S. Giner and M. S. Archer (London: Routledge & Kegan Paul, 1978), 197–227.

71. Caroline Hodges Persell and Peter W. Cookson Jr., "Microcomputers and Elite Boarding Schools: Educational Innovation and Social Reproduction," *Sociology of Education* 60, no. 2 (April 1987): 123–134.
72. Michèle Lamont and Anette Lareau, "Cultural Capital: Allusions, Gaps and Glissandos in Recent Theoretical Developments," *Sociological Theory* 6, no. 2 (Autumn 1988): 154–156.
73. Ibid., 155–156.
74. Margaret Archer, "Process Without System," in *Pierre Bourdieu*, ed. Derek Robbins (London: Sage, 2000), 1:118–143. *See esp.* 129.
75. Bonnie Erikson, "Culture, Class, and Connections," *American Journal of Sociology* 102, no. 1 (July 1996): 217.
76. Ibid.
77. Peter De Graaf, "The Impact of Financial and Cultural Resources on Educational Attainment in the Netherlands," *Sociology of Education* 59, no. 4 (October 1986): 237–246; Katherine McClelland, "Cumulative Disadvantage among the Highly Ambitious," *Sociology of Education* 63, no. 2 (April 1990): 102–121.
78. See chapter 4, "Is a Disinterested Act Possible?" Pierre Bourdieu, *Practical Reason*, 75–91.
79. Ibid., 77.
80. Ibid., 77.
81. Ibid., 77. Bourdieu uses the term *illusio* to describe the "fact of being caught up in and by the game, of believing the game is 'worth the candle,' or, more simply, that playing is worth the effort" (ibid., 76–77). In consistency with his prime metatheoretical objective, Bourdieu also defines illusio as "the enchanted relation to a game that is the product of a relation of ontological complicity between mental structures and the objective structures of social space" (ibid., 77).
82. Ibid., 78.
83. Ibid., 77.
84. Brubaker, "Rethinking Classical Theory," 109.
85. Chatterjee, *The Nation and Its Fragments*, 6.
86. Ibid., 5, emphasis in original.
87. Derek Robbins, "Introduction," in *Pierre Bourdieu*, ed. Derek Robbins (London: Sage, 2000), xv.
88. Ibid.
89. Ibid.
90. See Bose, "'Hindu Nationalism' and the Crisis of the Indian State," 104–164.

91. Ibid., 108–111.
92. Ibid., 108–111.
93. Calhoun, "Introduction," xi.
94. Ibid.
95. Bradford Verter, "Spiritual Capital: Theorizing Religion With Bourdieu Against Bourdieu," *Sociological Theory* 21, no. 2 (June 2001): 151.
96. Steven Engler, "Modern Times: Religion, Consecration and the State in Bourdieu," *Cultural Studies* 17, nos. 3–4 (May–July 2003), 445.
97. Ibid., 447.
98. Ibid., 445.
99. Ibid., 457.
100. Ibid., 454.
101. Verter, "Spiritual Capital," 150.
102. Ibid.
103. Ibid., 152.
104. Ibid., 152.

Chapter 3

The Emergence of Technological Capital in Colonial India

This chapter outlines key moments in the history of the technological field and technological capital and identifies important characteristics of the field and species of capital during the period of roughly 1750 to 1900.[1] Following Visvanathan, Baber conceptualizes the introduction of Western science and technology in India into three phases.[2] The first phase, following the British conquest of territories around the mid-eighteenth century, is the "era of the 'great surveys'...that witnessed the execution of a wide range of topographic, statistical, trigonometrical, cartographic and other surveys."[3] The second phase pertains to the introduction of technological education for Indians, culminating in a key moment in 1835 with a set of policy decisions by the British administration with regard to the education of Indians.[4] In the final phase, beginning roughly in the middle of the nineteenth century, the colonial state sought to institutionalize the relationship between technology and economy.[5] The role accorded for science and technology in the vision for the newly born Indian state at independence in 1947 derives from this long history.

Using this broad periodization as a frame to describe the relationship between technology and nationalism in the wider context of the colonial state and its relation to specific fields, I highlight some relevant changes in the economic and political fields in the early years of the colonial state. This is followed by a description of changes in the educational and cultural fields as the colonial state consolidated its presence. I next theorize the emergence of a relationship between nationalism and technology. The chapter concludes with a brief reflection on the relationship between technology and the economic field in the context of the emergence of "organized" anticolonial nationalism.

THE COLONIAL ENCOUNTER, THE EARLY COLONIAL STATE, AND THE ECONOMIC AND POLITICAL FIELDS

The precursor of the British colonial state in India was a private trading company: the British East India Company. British colonialism in India is approximately dated from the mid-eighteenth century, when the company established political control over Indian territories through conquest. Colonialism in its early phase amounted to a violent rupture of the Indian economic and political fields through the imposition of a nomos wrought elsewhere.[6]

Systems of trade among Indian communities had traditionally operated with a degree of autonomy from the operations of political power, reflecting "a traditional Indian pattern of a disjunction between trade and public functions" as well as the lofty disdain of Indian rulers toward "indigenous mercantile classes."[7] By the mid-eighteenth century, the imperial Mughal state had collapsed and had been replaced by autonomous "successor states."[8] But the logic of the political and economic fields did not converge. With regard to the political field, rulers of the successor states "everywhere sought legitimacy for themselves in religious terms."[9] Political power, in other words, was authorized as a special kind of religious capital. Norms of kingship articulated in the Mughal period endured in the successor states. Like their predecessors, the new rulers sought popular legitimacy by adapting to local cultural traditions. As a form of practical reason, the logic of the Indian political field had historically proposed an identity between a ruler and a territorial subject population, and there was no deviation from this pattern in the Indian successor states. The new rulers did not disturb the arrangements of religious patronage for local populations set in place under the Mughal empire. Habib observes that economic development continued more or less unhindered through the ascendancy of the successor states and the decline of the Mughal empire.[10] In medieval north India, then, the economic field was neither a model of Oriental despotism nor an arena of free exchange; while it possessed a degree of autonomy from the political field, it was also entangled with the complexities of social hierarchies and religious commitments.[11]

But with the company's political control of Bengal, the relationship between the political and economic fields would change in important ways. For a period of time, the nomos of the economic and political fields would merge. In 1765 the company was constituted as a *zamindar*—the possessor of a right to revenue from land—that controlled the *diwani* or revenue of Bengal, Bihar, and Orissa, but was also a trading entity. The combination of economic and political functions represented a radical shift from the traditional pattern of separation of political and economic power.[12] Habib observes that "the need for the Company to maximize what it obtained from grants of revenue was the single greatest motive for the British to become directly involved in the administration of Indian territory."[13] The dual status of the company as both political creature and economic entity lasted briefly.[14] But the interlude when the economic and political functions were merged witnessed a cynical exercise of political power by the company in the interests of profit. In 1770 Bengal suffered a famine. Although the chief cause of the famine may have been drought, its impact was strongly exacerbated by colonial policies.[15] Even a year after the famine, "the government insisted on increasing the revenue collection targets to make up for the lost revenue of 1770," leading to the impoverishment of a prosperous Indian region and peasant flight out of the region in search of alternative employment.[16]

The Permanent Settlement Act of 1793 embodied another change in the relationship between the economic and political fields. Zamindars, traditionally an intermediary group between the imperial state and peasant masses who were entitled to a share of revenue from land produce, were now reconstituted as private property holders who owed a share of earnings from the land to the British state. The colonial administration also hoped to procure the loyalty of the landholders to the British state by yoking together their economic interests with the political and economic interests of the company.[17] Rising mercantile classes in India also found land to be a valuable investment.[18] Cash-rich and prosperous merchants were often able to buy land from zamindars who had been unable to meet their financial obligations to the company. Land now came into the play of economic capital in new ways, and a kind of median aristocratic status based on ownership of land was authorized as political capital. By the mid-nineteenth century, along with the zamindari system, other initiatives relating to the institutionalization of private property, such as the *Ryotwari* settlement with "peasant proprietors" and *Mahalwari* settlement with the "village community," markedly impacted Indian systems of rights relating to land.[19] In each kind of settlement, the company's objective of maximizing revenue resulted in "arrears of payment, mounting debt, increasing land sales and dispossession."[20] The conflation of economic and political logic had implications for shifts in the nomos of the political field. Bandopadhyay notes that the impact of the changes may have been "less spectacular than once thought to be," but "the groups and classes that survived had substantially *different* rights, obligations, and powers" as reflected in agitations against British rule.[21]

Along with these shifts, the company's assertion of the right of exclusive sovereignty also proposed a new relationship between the political and economic domains. Technically under the sovereignty of the Mughal emperor, the company, however, proclaimed that it had rights of administration, governance, and control of law and order over its territories—functions that had been provided for by private individuals under the Mughal empire.[22] Guha, in fact, sees the moment of 1765 as signifying the birth of the colonial state, wherein the combination of economic and political functions was underscored by the invocation of law.

> At this point, in 1765, the East India Company's conquest of Bengal acquired legitimacy by a Mughal grant enabling it to collect the land revenues of the three Eastern provinces and to administer civil justice on behalf of the Nawab. Known technically as the Company's accession to Diwani—a technicality which has, alas, made some scholars insensitive to its significance to the truly inaugural moment of the Raj—it brought together in one single instance all three fundamental aspects of colonialism in our subcontinent, namely, its origin in an act of force, its exploitation of the primary produce of the land as the very basis of a colonial economy, and its need to give force and exploitation the appearance of legality.[23]

The entitlements of absolute sovereignty may not have been exercised immediately by the company.[24] But what is of relevance here is the new model of political authority suggested by the company's assertion and the conceptualization of a subject population implicit in that claim. The nomos, or logic, of the political field moved from the demand of a performance of nominal allegiance and tribute to a royal authority authorized in religious terms (as with the traditional model) to the demand of a performance of subjecthood and obligation to a secular power authorized by a sovereignty vested in the company. And, at least one way to contextualize the entry of Western science and technology into India is to understand it as consonant with the demand of this sovereignty for knowledge about its subjects.

Almost immediately in the wake of the acquisition of Indian territory by the British, "the issue of obtaining accurate information about the extent of the produce, the population, and other statistics about the area under their control acquired immediate significance."[25] British scientists, beginning with geographers, soldier-engineers, and botanists, would visit India to map the lands accurately for the purpose of drawing borders, enumerating populations, and gathering information about the ecological economies within which Indians lived and functioned.[26] Surveys inaugurated after the conquest and continuing through the colonial period "contributed to the reconstitution of India as a geographically defined region with precisely demarcated boundaries."[27] In the imagination of the British, India, hitherto "a nebulous entity, defined primarily in cultural terms," would now materialize in the full richness of her empirical reality.[28] India would be transformed from the heart of darkness to a "space constituted by technics," a land primed for the functioning of roads, railways, and telegraphs.[29]

The use of Western science and technology in the service of the post-conquest company can be seen as the first step toward the accumulation of "metacapital," that is, the accumulation of significant amounts of different species of capital by a nascent state toward the objective of constituting itself as a state proper.[30] In addition to economic capital, the nascent colonial state focused on accumulating informational, or symbolic, capital (through gathering "knowledge" about a subject population). The project of gathering knowledge was fundamentally related to the construction of statehood in the modern sense of the term. As one scholar puts it, one "symptom" of the modernity of the Indian nation-state as it was constituted in the course of colonial rule

> was that its techniques of government were very closely tied to techniques of measurement. From surveys of land and crop output to prospecting for minerals, from measuring Indian brains (on behalf of the false science of phrenology) to measuring Indian bodies, diets, and life spans (thus laying the foundations of physical anthropology and modern medicine in India), the British had the length and breadth of India, its history, culture, and society, mapped, classified, and quantified in detail that was nothing but precise even when it was wrongheaded.[31]

In this regard, the colonial state possessed a strategic advantage over its subjects since it could claim priority in defining what counted as capital (this strategic advantage, itself, may be understood as a characteristic of a state). Another consequence of the accumulation of metacapital by the colonial state may also be noted here. As Keay suggests, the tremendous human cost of the Great Trigonometrical Survey of India (GTS) and the activities related to the Great Indian Arc of the Meridian of the nineteenth century might have contributed to the tide of anti-British sentiment that culminated in the 1857 Rebellion, the first anticolonial uprising in India.[32]

With the foundation in 1784 of the Asiatic Society of Bengal by William Jones, the accumulation of scientific knowledge was linked to other economies and drawn into the operations of the cultural, educational, and political fields. "The institutionalization of Western science in India commences for all practical purposes with the establishment of the Asiatic Society of Bengal," but the society would also embark on the study of Indian customs.[33] This can be seen as an early step taken by the nascent colonial state in formalizing a structure of educational capital as well as symbolic capital, or prestige, honor, and recognition, that would accrue from membership in the society and/or recognition of the value of a researcher's work by the society.[34]

We might discern, here, the beginnings of a relationship between the political field, a newly introduced Western-style educational field, and an emergent technological field. Technological capital translated into political capital in that it served the purpose of governance, and it also translated, through another equation, into educational capital—specifically, knowledge about native peoples. Yet, Indians did not possess the language—English—in which one could participate in the exchange of capital in the Western-style educational or technological fields with even a limited degree of autonomy. For that, Indians would have to wait until a set of decisions taken by colonial administrators in 1835 would bring Indians into the structure of exchange of these species of capital. Before that, however, colonialism would severely undermine the bases of the indigenous educational "fields." What happened to culture was a more complicated, if no less intrusive, affair.

The Colonial State, Symbolic Violence, and the Cultural and Educational Fields

In its efforts to change the nomos of the cultural and educational fields, starting with a set of decisions in the late eighteenth century, the developing colonial state would commit the consummate act of symbolic violence on an indigenous realm of culture. The state would introduce reforms in the "laws" of Indians and would redefine the logic of community in primarily religious terms through legislation. It would also introduce scientific and technological education in English as well as the study of the English language and literature for Indians, while largely discontinuing patronage for indigenous education.

In the dense history of the development of colonial law, some crucial events may be noted. The long process of the codification of Hindu and Muslim laws as "personal laws" initiated by the British in 1772 takes center stage here. The homogenization of law essentially privileged textual sources of Hindu and Muslim law over customary law. Initially, the British relied on local priestly elites among Hindus and Muslims, *pandits* and *qazis*, to understand the law. This "set into motion the process of Brahmanization and Islamization of laws"—the inscription of purity in line with an idea of religious doctrine as interpreted by the priests.[35] The frustration of the colonial authorities with the priests and their suspicion of the motives of the latter soon led them to the decision to translate into English the texts from which Hindu and Muslim laws were derived.[36] This would result in the creation of the Anglo-Hindu law for Hindus and Anglo-Muhammadan law for Muslims in addition to personal laws for Christians. Jones would play an influential role in the process. Fluent in Persian and Arabic when he came to India, Jones would also master Sanskrit. Through translations of texts such as the *Manusmriti* and the *Al Sirajjiyah*, Jones codified into existence systems of Hindu and Muslim law.

In 1774 a Supreme Court was established in Calcutta, and in 1781 the Supreme Court was granted "express jurisdiction over natives."[37] The 1781 decision protected the rights of Hindus and Muslims to legislate "personal" matters, such as inheritance, marriage, succession, and rent, as per their native customs and laws. But there was an ambiguity in the colonial state's understanding of the scope of the personal and native. The conflation of customary practice and religious law resulted in

> a legal fiction that the laws of Hindus and Muslims are rooted in their respective scriptures and further that Muslims and Hindus are homogeneous communities following uniform laws...This legal fiction provided no space for validating the role of customary law which has no scriptural basis and is evolved at the local level transgressing boundaries of religious identities.[38]

By codifying separate personal laws for Hindus, Muslims, and Christians, the British administration signaled that in its imagination, these communities were essentially separate peoples whose customs were independent of each other. The constructs of "'religious laws' and 'personal laws' became interchangeable, but in the process it was forgotten that before the intervention of the British all other aspects of law for Hindus and Muslims were as religious as the personal laws."[39] Ironically, the British rationale of preserving the autonomy of communities violated the fact that Indian communities had traditionally been allowed to follow customs free from the interference of rulers.

For, despite the claim of noninterference, the colonial state had been interfering in the customs and religious affairs of communities well before embarking on the process of legal codification.[40] Even afterward, the British would introduce legal reforms in the realm of "personal" or "religious" affairs of communities. The Sati Regulation Act of 1829, which banned the Hindu practice of widows immolating themselves on their husband's funeral pyres, was a case in point. The British justified such legislation as legitimate exceptions to the self-proclaimed rule of noninterference on grounds of protecting the interests of Indian women.[41] Hindu reformers, notably Raja Ram Mohan Roy, were influential in the state's decision to ban sati. An emergent middle class in Bengal also supported the decision. Similar reformist acts would follow in the course of the century.

The colonial state would also reshape the logic of educational and cultural fields in the wake of the "Orientalist-Anglicist controversy" about the appropriate responsibilities of the colonial administration toward indigenous systems of learning and culture.[42] The Orientalist position proposed that patronage of indigenous education should be continued under the company, as had traditionally been the case under Indian rulers. The Anglicist position called for reform in the sphere of Indian education in the medium of the English language, though the Anglicists were divided about whether such intervention should be secular or whether it should focus on the promotion of Christian values. Zastoupil and Moir note that "despite the growing strength of the evangelicals, educational policy in India remained a minor concern of the East India Company until 1823 and then was largely set by the orientalists for another decade."[43] The ground for intervention, however, was progressively prepared over the early nineteenth century.

In 1835 Macaulay, a British administrator, put forth his "Minute on Indian Education"—a compendium of damning judgments on indigenous Indian knowledge as well as a set of specific policy recommendations—to Bentinck, then governor-general, as an argument for promoting Western education in India in the medium of English.[44]

Macaulay's framework for transforming Indian education, however, would not be Christianity. The vehicle for educational change, but more broadly and crucially cultural change, would be the English language. Education in the English language would serve a vital administrative and ideological function. It would create an intermediary class of Indians who could assist the British in governing the nation and could simultaneously be the instruments for suffusing Western cultural values among the Indian populace at large, in the process transforming the indigenous body politic with the lifeblood of Western reason. According to Macaulay, there was no virtue in patronizing indigenous education. If it were left to him, he would stop "the printing of Arabic and Sanscrit books… abolish the Madrassa and the Sanscrit college at Calcutta."[45] He would also abolish the stipend provided to students in indigenous educational institutions and divert the funds to institutions that promoted English learning.

The governor-general, Bentinck, broadly accepted Macaulay's recommendations with the Resolution of March 7, 1835. Based on protests against the recommended abolition of indigenous institutions of learning, the Resolution, however, refrained from abolishing such institutions, although it would not provide financial support for them. The decisions can be summed up as follows. The colonial administration would (a) support scientific and technological education for Indians in the medium of English; (b) found colleges for the promotion of such learning; and (c) for the most part, withdraw financial support for Indian institutions providing indigenous education in a number of areas including Sanskrit and Persian language and literature, medicine, astronomy, and mathematics.[46]

The decision of 1835 led to the decline of institutions of indigenous learning, with technology and the English language displacing the teaching and learning of Indian classical languages and Indian sciences. Not surprisingly, the strongest resistance to Western science and technology had come from Indian groups such as *vaids* and *hakims* (doctors) and *pandits* (learned Brahmins) whose social status was dependent on their traditional authority in the sphere of learning.[47] Whatever the motivation for their anxieties, the resistance of such groups turned out to be prescient. By the end of the nineteenth century, such institutions had practically vanished.[48]

Importantly, the benefits of Western technology were potentially available to a wider cross section of Indians beyond urban class and caste elites. One scholar has argued that "there was no resistance" to Western technology in Indian society "on the grounds of caste or creed."[49] Western technology might have represented the emancipatory possibility of challenging the authority of social hierarchies in India. As technological education helped produce Indian doctors and engineers from Indian elites, it also provided employment and training opportunities for Indians from relatively less privileged backgrounds, in industries such as shipping and shipbuilding, railways, and the electric telegraph. In these professions, Indians worked in subordinate capacities to their British superiors. We can conceptualize nascent "professional" classes emerging in India by the late nineteenth century, defined by occupation and not so much by caste or religion. In practice, however, during the colonial period, there was a "strong correlation between upper caste rank, English education, and the professional classes."[50] Traditional access to education on grounds of literacy as well as privileged caste backgrounds was an important factor in shaping these "Indian middle classes."[51] Brahmins and other literate segments of Indian society were at a distinct advantage over other social groups in exploiting the opportunities afforded by the policies of the colonial administration.

Macaulay's call for Western education in 1835 was supported by Indian reformers, such as Raja Ram Mohan Roy, who was strongly aligned with the Anglicist position in the education debate.[52] More generally, Indian urban elites that had emerged as a result of British policies of land reform in Bengal and elsewhere were overwhelmingly in favor of Western education. Even before the 1835 decisions, these "emergent middle classes" had founded several such institutions in the Calcutta and Bombay Presidencies and in Benares.[53] Schools and colleges that offered Western education in the medium of English (both prior to and after 1835) were quite popular among Indian elites.[54] Zastoupil and Moir point out that

> the number of students at Hindu College grew dramatically after 1835 as did those attending the new medical institution (with a decidedly western curriculum) which the British established about the time of Macaulay's minute; when a new English school (Hooghly College) was established by the GCPI in 1836, 1200 applicants appeared within three days for admission to the English department alone; over a two-year period the Calcutta book society sold more than 30,000 English books.[55]

In 1857, with the Universities Act, the universities of Calcutta, Bombay, and Madras were established. Punjab University in 1882 and Allahabad University in 1887 followed.[56]

Interestingly, despite the embrace of Western education in English by Indian elites, science and technical instruction appeared to be among the least preferred subjects of instruction among Indians. Aggarwal provides a profile of Indian colleges by subject and the student population for each type of college in 1901–1902.[57] In 1901–02 the total number of colleges in India was 191 and the total number of students was 23,909. The overwhelming majority of these colleges were Arts Colleges–English (providing humanities instruction in English), numbering 140 with a total of 17,048 students. Arts Colleges–Oriental (providing education in Indian languages) numbered only 5 with 503 students. Under the category of Professional Colleges are included Law, Medicine, Engineering, Teaching, and Agriculture. Here, Law held the top spot with 30 colleges and 2,767 students, but was a poor second when compared to Arts Colleges–English. Medicine and Engineering colleges numbered only 4 each, with 1,466 and 865 students, respectively. Teaching numbered 5 colleges and had a total of 190 students, and Agriculture numbered 3 colleges and a mere 70 students. Technical education, under which we may count Medicine, Engineering, and Agriculture, accounted for just about 10 percent of all Indian students, while Arts Colleges–English education as a whole accounted for a staggering 73 percent of all Indian students.

Aggarwal does not provide a separate category of Science Colleges–English. McDonald, in an essay on higher education in Bombay in the late nineteenth century, reveals that instruction in science subjects in the medium of English was provided by colleges also offering "Arts" subjects.[58] But Tangri notes that because of factors such as the cost of equipment, access to science classes was extremely limited.[59] In the 1897–1902 period, college education was "excessively literary" in character, and "less than one in a hundred successful candidates obtained a science degree."[60] For the 6,790 who obtained a Bachelor of Arts degree, a mere 54 obtained a Bachelor of Science degree.[61]

The differential in the number of students in Arts Colleges–English and, broadly, colleges specializing in science and technology may be a function of the number of colleges respectively offering these types of education, as well as factors such as cost. Yet it might also be a function of demand-and-supply for these types of education among Indian elites, for the early history of English-medium education in India suggests that the British partially justified the 1835 decisions on the perceived demand for instruction in English, which seems borne out in reality by the actions of Indian elites. We might see here some important features of the educational and technological fields in India. The nomos of the educational field arguably reflects the imperative of acquiring prestige, which, at this point of time, rests with the English language as the signifier of culture as much as it might with any scientific or technological discipline as such. The nomos of the technological field may also be conceptualized as a combination of scientificity and technicity, where scientificity signifies the values of reason and rationality and technicity the pragmatic application of these values in concrete material contexts. With these observations as a backdrop, I now turn to an assessment of the relationship between technology and elite nationalism in colonial India.

THE RELATIONSHIP OF TECHNOLOGY TO ELITE NATIONALISM

Arguing that the theoretical category of "religion" may be limited in its analytical value for assessing Indian society, Chakrabarty states that "the practices that we gather under the name *religion* do not repeat the history of the Enlightenment. Such translations are by definition hybrid and incomplete."[62] Indeed, the secularization thesis, or the separation of church and state in the European context, which appears fairly central for Bourdieu's understanding of the religious field as a sphere of the operations of organized religion, has no analogue in Indian history.[63] Political power in precolonial India was secular in its operations yet was legitimated in terms of religious authority. While one can make a case that there existed a pragmatic arrangement between different communities reflective of the values associated with modern-day secularism, this state of affairs did not necessarily imply the relegation of religious sentiment to the private sphere.

However, conflating the sphere of culture and religion into the notion of a cultural-religious field as a theoretical resolution may not be very helpful either. Ray's insightful reading, which traces the development of notions of Hindu and Muslim identity through Indian history, exemplifies this point.[64] Ray suggests that Hindu and Muslim identities possessed a respective internal coherence that emerged as the result of a complex interaction between the two, involving cross-fertilization as well as conflict.

Hindus and Muslims saw themselves and each other as two peoples or "nations" (*qaum*), yet both were "Hindis," that is, part of a larger collective of the territory of "Hind." Ray offers a valuable theorization of Hindus and Muslims as "communities of sentiment" that articulated themselves as different yet sharing important cultural commonalities.

In view of this, the concepts of field and capital appear somewhat lacking for reading the relationship between religion and culture in early colonial India. Bourdieu's conceptualization of the state may offer a way out. As noted earlier, Bourdieu posits that the perceptual categories of the citizens of a state are historically constituted through a shared relationship between state and people. This incorporation of citizens within the state and, correspondingly, of the state within the citizens explains why citizens are disposed to "comply" with the state's dictates. But for the outcome of that struggle between state and people to be legitimately accepted by society, there must be the premise of equality or of fairness: the illusion that all those engaged in the struggle for capital might be able to shape the outcome of the field.[65] Here, we can identify a defining characteristic of the colonial state: the colonial state is never fully incorporated within its Indian subjects and the colonial state's categories of thought are not shared by its subjects in precolonial and colonial India, with regard to "religion" and culture.[66] Accordingly, the colonial state's actions in the realm of the laws, practices, and customs of Indians—the cultural and religious fields—may be read as the establishment of a framework by the state where subjects could initially be exposed to state categories of thought, which emphasize the civilizational superiority of the British and their right to govern India. The colonial state's decision to reconstitute the educational fields through the medium of English education in 1835 can also be seen as a project to inculcate state categories of thought in Indians.

One can also see the post-1835 cultural, educational, and religious fields in India as unstable formulations, where the nomos of the fields is conceptualized somewhat differently by the British state and Indians, even if there is some consonance in their understandings. The categories of thought proposed by the colonial state are neither completely rejected nor completely accepted; however, the fundamental claim corresponding to the nomos of the field as defined by the British—the right to govern Indians—is never fully accepted by all Indians as the fundamental rule of the game, even if sections of Indian elites agree with that position. The anticolonial nationalist movement can consequently be read as the struggle to progressively redefine the nomos of the fields contested between Indians and the British. We might, then, meaningfully assess the zone of interaction between the British and Indians in the various fields, as these were sought to be redefined by British policies. It is the fault line in the nomos of these overlapping fields, with regard to their perception by the colonial state and Indians, that enables the articulation of a relationship between technology and anticolonial nationalism. Technology as scientificity played a key role in the anticolonial nationalist imagination, by authorizing it as a distinct species of capital that also translates into cultural capital.

As Baber notes, it was only after 1835, when the colonial administration extended its patronage to scientific education in English, that Indians were granted the opportunity to prove their scientific abilities.[67] Prakash, in a brilliant reading, argues that the British decision to extend scientific learning to Indians embodied a paradox.[68] On the one hand, the decision reaffirmed the discourse of civilizational difference between Europeans and non-Europeans, which defined the European West as rational, enlightened, and civilized and the non-West as primitive and irrational. On the other hand, it was an indirect admission by the British that Indians were capable of rational scientific thought. Indians could stake a claim to a universal humanity and rationality via the medium of science, since science stood for universal and rational values.[69] Scientificity as the embodiment of reason and rationality was authorized as a kind of universal cultural capital that was not contingent upon race or nationality, although in colonial discourse, race itself endured as a non-accumulable mode of cultural "capital."

Prakash suggests that this paradox set in place a problem for the Indian elite—the reconciliation of one's universalism with the realities of subjugation under the colonial power—but also outlined a manner of thinking national identity among Indian anticolonial nationalists.[70] The work of Orientalist scholars, available to Indians through the exercise of the colonial state in accumulating cultural metacapital through initiatives such as the Asiatic Society and surveys, offered an educated intelligentsia the basis for a conceptual resolution of the problem. The Orientalist thesis proposed that ancient India was an advanced, prosperous, and learned civilization that over the centuries had been victimized by waves of invasions and, as a result, had fallen into a pitiable state of decay. Drawing on the thesis, a predominantly educated English-speaking Hindu elite, encompassing reformers, historians, writers, and scientists, constructed a narrative of Indianness through an interpretation of Hindu texts. In this reading, Hinduism was described as a universal scientific, religious, and cultural worldview—a "Hindu science" that was the foundation of an authentic Indian identity as well as the source for the realization of that identity.[71] However, if this science had manifested itself in all its fullness in an ancient time, Indian civilization had been reduced to backwardness, superstition, and ignorance because of the savagery inflicted by Islamic invaders and the economic exploitation caused by British rule. This reinterpretation of Hinduism, on the one hand, offered a basis to challenge the hegemony of Western claims over reason, since, the argument went, Hindu thought had been scientific since its very inception, predating by centuries the European Enlightenment.[72] On the other hand, in its particular reading of the Vedas, which entailed a rejection, for instance, of the "sensual playfulness" of the Puranas, it refashioned Hinduism along the lines of an austere, monotheistic, and Sanskritic "religion."[73] Simultaneously, it was characterized by a communal and anti-Muslim consciousness. What is central to the logic of this reappropriation of the Indian heritage, Prakash argues, is that the past is constructed as an "anteriority"—something that is not irretrievably lost to time but an always accessible reservoir of cultural and scientific resources that can be grasped through an enlightened recognition of the truth of what it means to be Indian.[74]

Scientific knowledge and learning in colonial India operated, in this manner, as an arena in which the supremacy of the West could be contested.

This contestation meant accepting, in one critical respect, the categories of classification proposed by the colonial state with regard to the technological field. The Indian elite did accept the self-definition of science as synonymous with rationality and enlightenment. At the same time, the contestation meant rejecting in one important respect the nomos of the field as proposed by the British, for the Indian elite, through making a case for a "Hindu" science, did not accept the self-definition of science as inherently or exclusively Western in its origins or character. In the Orientalist reading, the glory of Hindu India was lost to the past or, at the very least, would require the agency of the British to resuscitate it; in the Indian rereading of the Orientalist narrative, the pastness of the Hindu past could be reactivated through the cultivation of the scientific temper or scientificity. Colonial ideology proposed that scientificity was an embodiment of Western cultural capital. In contrast, the Indian elites understood scientificity as both the embodiment of and an instrument for the ready accumulation of a Hindu cultural capital whose value exceeded that of Western cultural capital.

Indeed, inasmuch as science stood for the intrinsic cultural superiority of the British colonizers or European civilization, that proposition was expressly rejected by the Indian nationalist elite. Chatterjee makes a strong case for the total autonomy of an Indian cultural field from the framework of state reason in the colonial era.[75] Chatterjee argues that anticolonial nationalism visualizes an autonomous "domain of sovereignty within colonial society well before it begins its political battle with the imperial power."[76] He describes this as a separation between a realm of cultural or spiritual values, an "inner" domain where those colonized hold authority, and the "domain of the outside," the sphere of "economy and statecraft, of science and technology," where the West has vanquished the natives.[77] Anticolonial nationalism, accordingly, defined its task as one of wresting back control of the outer domain, while it could rest assured that its cultural identity was uncontaminated by the colonial encounter.[78]

In tandem with Prakash's arguments previously described, Chatterjee's theorization allows us to conceptualize the nature of a further development in the nomos of the technological and cultural fields as understood by Indians. The Indian anticolonial nationalists conceded to the West a superiority in technicity, as belonging to the "outer" domain, where the West held an advantage over Indians. Yet scientificity—as a cultural value representing reason and enlightenment—belonged to the inner domain of Indian culture and here the Indians did not concede superiority to the West. There are two implications of this development in the nomos of the technological and cultural fields, respectively. First, the "outer" domain was conceived of as culturally value-neutral; hence, technicity was constructed as value-neutral. Second, scientificity as a cultural value would be assimilated in the "inner" domain as one of several Indian cultural values. Although scientificity as rationality and reason would be invoked by the Indian nationalist elite as the basis for reform in cultural practices, scientificity would by no means erase those "religious" and cultural practices—of Hindu dharma—that did not square with the dictates of European reason. However, with regard to technicity as well, the Indian nationalists would challenge the nomos of the technological and economic fields as defined by the British.

Technology, the Economic Field, and "Official" Anticolonial Nationalism

In the third phase of the introduction of Western science and technology in India, the colonial state sought to systematize and institutionalize relations between science, technology, and the economy.[79] Western technology as technicity manifested itself in the form of an industrial apparatus to support a colonial economy, and technological capital of this variant translated into economic capital. This aspect of the nomos of the technological field was accepted by the Indian nationalist elites. The anticolonial nationalist elites did not dispute that industrialization and the technological modernization of the Indian economy were the means to economic progress. What they disputed, however, was the British claim that the colonial state could best serve the economic field. The Indian nationalist elites argued that the nomos of the political and economic fields, as defined by the British, were severely in conflict with each other and that Indians themselves would need to control the political field in order for economic progress to suffuse across Indian society at large. This was one of the main issues on which organized anticolonial resistance, from its inception through the time of Indian independence, would challenge the British domination of India.

A convenient date for the start of organized anticolonial political resistance in Indian society is 1885, when India's first nationalist party, the Indian National Congress, was founded.[80] The leadership of the Indian National Congress was largely composed of upper-caste, Hindu, educated, English-speaking Indians, and support for the Indian National Congress was predominantly provided by urban Hindu elites.[81] The nationalists in the Congress argued that British policies favored European businessmen over Indians, that British rule had resulted in the economic exploitation of Indian society and the impoverishment of her masses, and that Britain could not reconcile its interests as an imperial power with the economic well-being of Indian society. Dadabhai Naoroji's "drain of wealth" thesis helped forge an economic nationalism that would endure through the period of organized anticolonial nationalism. Corbridge and Harriss point out that right through the 1880s to the 1940s when India achieved independence, one of the chief issues of contestation between the nationalists and the British administration was the economic exploitation of India.[82] The British administrators, it is worth noting, strongly rejected the "drain of wealth" thesis.[83]

Marxist, Nationalist, and Left-Nationalist Indian scholars generally appear to accept the argument that British colonialism was inherently exploitative of Indians in economic terms. There is much evidence to support this claim. As Davis notes,

> If the history of British rule in India were to be condensed into a single fact, it is this: there was no increase in India's per capita income from 1757 to 1947. Indeed, in the last half of the nineteenth century, income probably declined by more than 50 percent.[84]

Recently, however, scholars have questioned the extent to which India as a colony was an economically beneficial proposition for the British state and whether the impoverishment of Indians under colonial rule can be taken as a legitimate economic variable.[85] The intricacies of these debates are not relevant here. The important fact to note is that the Indian nationalists and the colonial state disagreed strongly on the issue of the control of the benefits of technicity, although they agreed on the more fundamental proposition that technicity itself was an instrument of economic progress. As was the case with scientificity and its relationship to the cultural field, on the matter of technicity and its relationship to the economic field, too, the Indian nationalists did not entirely share the categories of state thought with the British colonizers after the very early stages of anticolonial nationalism.

The question of whether technicity as a species of technological capital continues to be constructed as value-neutral gets somewhat complicated here, for one can read the Indian nationalist argument that technicity should benefit the Indian populace at large as the beginnings of a "new" cultural value of Indian modernity. One can trace the glimmerings of a mode of imagining the nation within which technicity is subsumed as part of an Indian culture that is at once timeless and modern and, at once, materially beneficial and socially just. Economic nationalist arguments alone would not win India her independence in 1947. But questions of scientificity and technicity would permeate the realm of economic and cultural discussions, the debates on Indian and Western values, and the contests over the character of the Indian state. Gandhi and Nehru would propose radically different understandings of scientificity and technicity, the relationship of technology to Indian and universal values, and the role that should be accorded to technology in the independent Indian state. A third figure, Veer Savarkar, the founder of the ideology of Hindutva, or Hindu nationalism, would offer another, alternative vision of India that would, in part, be founded upon a response to these very issues. As with much of Indian history, the question of who won or lost in the long run is not as easy to answer as it may seem at first sight.

ENDNOTES

1. Technological capital is different from Bourdieu's definition of scientific capital, which is a specialized surplus form of cultural capital bestowed upon scientists recognized by their peers as having made an outstanding contribution. Technological capital may be accumulated by a range of participants including, but not limited to, scientists.
2. Baber, *The Science of Empire*, 137–138; Visvanathan, *Organizing for Science*, 8–14.
3. Baber, *The Science of Empire*, 137–138.
4. Ibid., 137.
5. Ibid., 138.
6. My arguments here draw significantly on the work of historians of colonial India. For Guha's argument about the inherent alienness of colonialism, see chapter 1, "Colonialism in South Asia: A Dominance without Hegemony and Its Historiography," in *Dominance without Hegemony: History and Power in Colonial India* (Cambridge, MA: Harvard University Press, 1997), 1–99. Also see Guha's *An Indian Historiography of India: A Nineteenth-Century Agenda and Its Implications* (Calcutta, India: Center for Studies in Social Sciences, 1988). Describing the "economic disruption of India," Habib states that India's economic history takes on a special importance from the mid-eighteenth century onward: "subjugated by the first industrial nation of the world, it offered the classic case of the colonial remoulding of a pre-colonial economy." Irfan Habib, "Studying a Colonial Economy—Without Perceiving Colonialism," *Modern Asian Studies* 19, no. 3 (1985), 355, 359. Also see Irfan Habib, "Introduction," in *The Eighteenth Century in Indian History. Evolution or Revolution?* ed. P. J. Marshall (New Delhi: Oxford University Press, 2003), 1–49. Sudipta Sen in his work describes the East India Company as a colonizing state that had begun preparing the ground well before the "formal dates of conquest," foreshadowing its transition to a fiscal-military state. See Sudipta Sen, *Empire of Free Trade: The East India Company and the Making of the Colonial Marketplace* (Philadelphia: University of Pennsylvania Press, 1998), 18.
7. Deepak Lal, *Cultural Stability and Economic Stagnation: India c1500 BC–AD 1980* (Oxford, U.K.: Clarendon Press, 1988), 106.
8. Habib, "Introduction," 3.
9. Ibid., 8.

10. Ibid., 15.
11. Sen, *Empire of Free Trade*, 19–21.
12. Deepak Lal, *Cultural Stability and Economic Stagnation*, 106.
13. Habib, "Introduction," 29.
14. Deepak Lal, *Cultural Stability and Economic Stagnation*, 106–107. With Cornwallis' restructuring of the company, "political power was once again to be strictly separated from economic power" (107).
15. Baber, *The Science of Empire*, 161.
16. Ibid., 162.
17. B. L. Grover and S. Grover, *A New Look at Modern Indian History: From 1707 to the Modern Times* (Delhi: S. Chand, 1988), 87.
18. Deepak Lal, *Cultural Stability and Economic Stagnation*, 107.
19. Sekhar Bandopadhyay, *From Plassey to Partition: A History of Modern India* (Delhi: Orient Longman, 2004), 95.
20. Ibid.
21. Ibid. (emphasis added).
22. Habib, "Introduction," 30.
23. Guha, *An Indian Historiography of India*, 4. Guha reads the political field here as structured by a combination of force and law. Sovereignty may correspondingly be understood as the quality sought to be protected and affirmed through law.
24. Habib, "Introduction," 27.
25. Baber, *The Science of Empire*, 138.
26. This also served the development of European scientific knowledge itself. See Baber, *The Science of Empire*, 151–153.
27. Ibid., 151.
28. Ibid., 151.
29. Prakash, *Another Reason*, 11.
30. Bourdieu, *Pascalian Meditations*, 40–41.
31. Dipesh Chakrabarty, *Habitations of Modernity: Essays in the Wake of Subaltern Studies* (Chicago: University of Chicago Press, 2003), 83–84.
32. Keay, *The Great Arc*, 169–170.
33. Visvanathan, *Organizing for Science*, 8, quoted in Baber, *The Science of Empire*, 155; Baber, *The Science of Empire*, 155.
34. We can also understand symbolic capital as a kind of universally applicable capital that derives from the accumulation of a critical mass of other kinds of capital in any given social setting. Association with the Asiatic Society would have provided symbolic capital in the form of prestige and status. It would also have indicated possession of a critical mass of educational capital.

35. Flavia Agnes, *Law and Gender Inequality: The Politics of Women's Rights in India* (New Delhi: Oxford University Press, 1999), 44.
36. Ibid.
37. Ibid., 43.
38. Ibid., 43.
39. Archana Parashar, *Women and Family Law Reform in India* (New Delhi: Sage Publishers, 1992), 66; quoted in Janaki Nair, *Women and Law in Colonial India: A Social History* (New Delhi: Kali for Women, 1996), 181.
40. Nair, *Women and Law in Colonial India*, 181.
41. Agnes, *Law and Gender Inequality*, 47. In this regard, Agnes points out that "an image of the cruel and superstitious natives who needed Christian salvation was deliberately constructed by the Evangelists" (46).
42. For a good summary of this debate, see Lynn Zastoupil and Martin Moir, "Introduction," in *The Great Indian Education Debate: Documents Relating to the Orientalist-Anglicist Controversy 1781–1843*, ed. Lynn Zastoupil and Martin Moir (Richmond, U.K.: Curzon, 1999), 1–72.
43. Ibid., 8.
44. Thomas Babington Macaulay, "Minute on Indian Education," in *Archives of Empire Volume 1: From the East India Company to the Suez Canal*, ed. Mia Carter and Barbara Harlow (Durham, NC: Duke University Press, 2003), 227–238. Indian knowledge, he wrote, was "false history, false astronomy, false medicine…in company with a false religion" (236).
45. Ibid., 238.
46. In 1839 a limited amount of funds were sanctioned for institutions providing indigenous education. But Wood's educational dispatch in 1854 clearly underscored the administration's prioritization of European knowledge. Baber, *The Science of Empire*, 205.
47. Sangwan, *Science, Technology and Colonisation*, 135.
48. Baber, *The Science of Empire*, 225.
49. Sangwan, *Science, Technology and Colonisation*, 133.
50. Francine Frankel, "Middle Classes and Castes in India's Politics: Prospects for Political Accommodation," in *India's Democracy: An Analysis of Changing State-Society Relations*, ed. Atul Kohli (Princeton: Princeton University Press, 1988), 226.
51. Ibid., 225.
52. Macaulay, in fact, borrowed from Roy his arguments about the worthlessness of Sanskrit as a subject of learning for Indians. Zastoupil and Moir, "Introduction," 35.
53. Baber, *The Science of Empire*, 195.
54. Zastoupil and Moir, "Introduction," 30.

55. Ibid.
56. J. C. Aggarwal, *Documents on Higher Education in India (1781 to 1992): Selected Educational Statistics: World Overview* (Delhi: Doaba House, 1993), 19. Aggarwal points out that the universities "were mere affiliating and examining bodies" and that they did not do any "direct teaching work" (ibid.).
57. Ibid.
58. Ellen E. McDonald, "English Education and Social Reform in Late Nineteenth Century Bombay: A Case Study in the Transmission of a Cultural Ideal," *The Journal of Asian Studies* 25, no. 3 (1966): 453–470.
59. Shanti S. Tangri, "Intellectuals and Society in Nineteenth-Century India," *Comparative Studies in Society and History* 3, no. 4 (1961): 372.
60. Ibid., 372.
61. Ibid.
62. Chakrabarty, *Habitations of Modernity*, 28, emphasis in original.
63. Verter's substitution of religious capital with "spiritual capital" as a particular kind of cultural capital can be helpful for assessments of the colonial period, but its value for the precolonial period may be more limited. See Verter, "Spiritual Capital," 150.
64. Rajat Kanta Ray, *The Felt Community: Commonality and Mentality Before the Emergence of Indian Nationalism* (Delhi: Oxford University Press, 2003).
65. Bourdieu, *Practical Reason*, 77.
66. This claim is consistent with Guha's Gramscian articulation of "dominance without hegemony" as the defining characteristic of the colonial state. See chapter 1, "Colonialism in South Asia: A Dominance without Hegemony and its Historiography," in Guha's *Dominance without Hegemony*, 1–99.
67. Baber, *The Science of Empire*, 138.
68. Prakash, *Another Reason*, 8–9, 46–48.
69. Ibid., 8. Michael Adas has argued that this is what distinguishes the introduction of Western science and technology in colonial India from colonial Senegal. Michael Adas, "Scientific Standards and Colonial Education in British India and French Senegal," in *Science, Medicine and Cultural Imperialism*, ed. Teresa Meade and Mark Walker (New York: St. Martin's, 1991), 4–35. It is true that in the case of both India and Senegal, a perceived lack on the part of the native subjects was attributed in part to the "enervating climate and centuries of barbarism" (18). And British educational policy in India was also marked by the logic of racial difference that was central to the European colonial project. However, the mode in which racism was institutionalized as a reason of educational policy in each case suggests that French colonialism deemed the Senegalese racially incapable

of learning Western science whereas British colonialism conceded the capacity for scientific understanding to Indians.

70. Prakash, *Another Reason*, 8–9. See also 98–120.
71. Ibid., 9. Prakash offers a detailed treatment of "Hindu science" in his chapter 3, "The Image of the Archaic," 86–120.
72. For our purposes here, it is the ideological mobilization of the idea of Indian scientificity as the basis of anticolonial nationalism that is of interest. In *The Science of Empire*, Baber describes ancient Indian innovations in mathematics, astronomy, and medicine in some detail. See chapter 2, "Science, Technology, and Social Structure in Ancient India," 14–52.
73. Prakash, *Another Reason*, 9.
74. Ibid., 91.
75. Chatterjee, *The Nation and Its Fragments*, 6. Prakash is in agreement with Chatterjee's understanding of the character of anticolonial nationalism. However, Prakash argues that Chatterjee fails to make a conceptual distinction between the idea of the Indian nation in the anticolonial struggle and the Indian nation-state which came to fruition as a result of this struggle. See Prakash, *Another Reason*, 201–203.
76. Chatterjee, *The Nation and Its Fragments*, 6.
77. Ibid.
78. Chatterjee argues that the Indian conceptualization of "inner" and "outer" domains did not correspond to the divide between the "private" and "public" sphere characteristic of European societies (ibid.).
79. Baber, *The Science of Empire*, 138.
80. As suggested in the earlier section, an anticolonial nationalist elite had emerged well before this date. Also, a number of nineteenth-century movements for social and religious reform, such as the Arya Samaj and Brahmo Samaj, often defined their agenda as a critique of aspects of colonial rule. The founding of the Indian National Congress in 1885 is often used as an originary moment in conventional nationalist histories. Here, it is used as a convenient marker to define some aspects of early "official" Indian nationalism.
81. Corbridge and Harriss, *Reinventing India*, 14, 242 n. 32.
82. Ibid., 14.
83. John McLane, "The Drain of Wealth and Indian Nationalism at the Turn of the Century," in *India and the World Economy 1850–1950*, ed. G. Balachandran (Delhi: Oxford University Press, 2003), 70–71.
84. Mike Davis, *Late Victorian Holocausts: El Niño Famines and the Making of the Third World* (New York: Verso, 2001), 311.
85. See, for instance, the essays in G. Balachandran, ed., *India and the World Economy* (Delhi: Oxford University Press, 2003).

CHAPTER 4

DESIGNING A TECHNOLOGICAL FIELD FOR AN INDEPENDENT NATION

This chapter traces the development of the technological field and capital in India over the course of the better part of the twentieth century. I approach the relationship between technology and nationalism in the phase prior to independence in 1947 through an assessment of three competing visions for the Indian nation-state, each of which offers a philosophical perspective about the nature of the state and proposes a corresponding role for technological development as a matter of state responsibility. These visions are, respectively, the following: the decentralized, industry-free, Indian nation proposed by Gandhi, premised on a radical critique of Western industrial modernity and the very idea of the nation-state; the dream of a modern Indian nation-state that would derive its strength from its essentially Hindu core, proposed by Savarkar, the main ideologue of Hindutva, or Hindu nationalism; and the Nehruvian socialist model of a technologically developed, economically prosperous, and socially progressive Indian nation-state.[1] The chapter also addresses the role of technology in the era of the Nehruvian socialist state from 1947 to 1991.[2] The Nehruvian paradigm yoked ambitious projects of scientific and technological development with a state vision of socialism and secularism in the hope of achieving economic prosperity and social justice for all segments of Indian society. I describe the dynamics of exchange between the technological, economic, and cultural fields and these species of capital. The chapter concludes with a description of the characteristics of a technological class, in Bourdieu's sense of the term, in independent India.

THREE VISIONS OF THE MODERN INDIAN NATION-STATE

Each of the perspectives of Gandhi, Nehru, and Savarkar may be understood as a response to the problem of a nationalism seeking to transition from the anticolonial to the postcolonial: how were the political institutions that were the legacy of European rule to be endowed with a distinctive national character? In the answer that each provided, Nehru and Gandhi, and to a lesser extent Savarkar, explicitly addressed the role of technology in such a transformation.

Gandhi: An India of Autonomous Village Republics

Of the three nationalist figures, Gandhi was the strongest critic of the very idea of the nation-state and an equally trenchant critic of science and technology, both of which he saw as misguided aspects of Western modernity.[3] Gandhi rejected the path on which Western civilization had embarked as a particularly misguided trajectory of modernity. Neither the nation-state nor the instruments of material progress, such as science and technology, were viable for any society—Indian or otherwise. In *Hind Swaraj*, written in 1909, Gandhi presented a scathing criticism of Western civilization, which is partially predicated on a critique of both the idea of the modern state and science and technology.[4]

Gandhi saw the state as a coercive, centralized, and violent agglomeration of power.[5] An amoral abstraction, the state severed the individuals empowered to act in its name (such as civil servants) from their obligations of responsibility to their fellow human beings. The state was not just indifferent to the human but was antihuman. It sought to reduce the human being from a complex moral, political, and psychological entity to the homogenized abstraction called the citizen, who could be conditioned to blindly obey its dictates. The European model of the nation-state was especially unsuitable for India. The materialist, depersonalized, violent, and conformity-demanding character of the state was in direct opposition to the spiritual, concretely embodied, tolerant, and pluralistic character of Indian society. Gandhi feared that with independence the Indian nation-state would be controlled by an "urbanised and Westernised elite alienated from ordinary people" and would be as predatory as the British colonial state that it had replaced.[6]

Gandhi's views on science and technology mirrored and were consistent with his views on the state. Parekh suggests that Gandhi's disagreement with the idea of the state was located within a broader objection to science itself.[7] These objections, in turn, were closely related to Gandhi's philosophical understanding of human existence.[8] For Gandhi, human existence was irreducible to any one dimension of existence—for instance, the economic. In the intricate rhythms of human life, the economic was interwoven with the ecological and social, the spirituality of the individual was inseparable from the morality of the collective, and man's relationship with nature was linked to the moral-spiritual framework of everyday life.

Crucially, for Gandhi, social life was contingent on human particularities, which we may understand as human limits; the essence of the human, partly at least, derived from the physical limitations of the body, around which the rhythms of both individual and collective life were organized. Nandy suggests that it is the limitedness of the body that enabled Gandhi to link the human body and the body politic: "Bodily politics defines not only the nature of activity, but prescribes its limits."[9] The human, so defined, is the source of personal, as well as communal, spirituality. It is here that we find the heart of Gandhi's critique: science and technology subjected humans to a force far beyond the control of human limitations, a force whose impact would be devastating for *all* aspects of individual and collective life.[10] The example of the railways stood as testimony to the destructive consequences of the reengineering of social life through the agency of technology, beyond the natural limits of the human. "Man is so made by nature," Gandhi argues in *Hind Swaraj*, with reference to the invention of the railways,

> as to require him to restrict his movements as far as his hands and feet will take him...God set a limit to man's locomotive ambition in the construction of his body. Man immediately proceeded to discover means of overriding the limit.[11]

The Faustian bargain, typified by the embrace of the railways by society, had cost man dearly, resulting, as it did, in a profound moral disorientation. "I am so constructed," Gandhi goes on to say of the moral cost of such overreaching,

> that I can only serve my immediate neighbors, but, in my conceit, I pretend to have discovered that I must with my body serve every individual in the Universe. In thus attempting the impossible, man comes in contact with different natures, different religions, and is utterly confounded.[12]

In this manner, the "most dangerous institution of the railways," Gandhi asserts, has led man "further away from his Maker."[13] But the toll taken by the railways was not restricted to psychological disorder and moral chaos. In a fascinating passage in *Hind Swaraj*, Gandhi described in concrete terms the havoc caused to Indian society by the railways.

> It must be manifest to you that, but for the railways, the English could not have a hold on India as they have. The railways, too, have spread the bubonic plague. Without them, masses could not move from place to place. They are the carriers of plague germs. Formerly we had natural segregation. Railways have also increased the frequency of famines, because, owing to facility of

> means of locomotion, people sell out their grain, and it is sent to the dearest markets. People become careless, and so the pressure of famine increases. They accentuate the evil nature of man. Bad men fulfil their evil designs with greater rapidity. The holy places of India have become unholy. Formerly, people went to these places with great difficulty. Generally, therefore, only the real devotees visited such places. Nowadays, rogues visit them in order to practice their roguery.[14]

The passage is astonishing in its condemnation of the railways as a negative force on multiple grounds. The railways are a political force in enabling the British to control India; they are an economic force in subjecting traditional local economic systems to the violence of market demand and supply; they are an ecological force as carriers of disease; and they are a religious force, resulting in the erosion of moral determination by rendering easy the accumulation of karma through pilgrimage. In each of these arguments, it is the impact of the railways on the natural condition of the human that is Gandhi's target. As Sahasrabudhey notes with regard to Gandhi's argument in *Hind Swaraj*,

> Machinery or the excessive use of machinery is rejected on the basis that it degrades and enslaves man. Not once is it mentioned that the excessive use of machinery is likely to cause a severe unemployment problem in the country.[15]

Gandhi's objection to technology is never purely on economic grounds; his economic arguments are ultimately grounded in moral categories. In evaluating the costs of technology and industrialization in the West, Gandhi similarly bemoaned the loss caused by the transgression of human limits through technology.[16] The moral logic invoked by Gandhi is not restricted to its Indian provenance but is formulated as a universalist rationale.

But Gandhi believed that the foundations of Indian society had not been irreparably damaged despite the violence of the encounter with Western modernity. Envisioning an alternative political modernity for India, he presented an idea of the country as a horizontal multitiered structure of decentralized, relatively autonomous, and self-sustaining communities, from the village through the district all the way outward to the government. In this collective of collectives, power would be localized and dispersed, in contrast to the coercive and centralized top-down structure of the modern nation-state.

A rejection of both capitalism and communism, the basis of life in Gandhi's India would be moral and spiritual.[17] The forms of natural human existence would take concrete shape in village communities. The provision of autonomy of social life at the village level, district level, or wider levels all the way to the largest collective of the nation would allow for individuals and communities to develop their spiritual lives in the context of their material lives, unhindered by either other communities or larger concatenations of power such as the state or any of its bureaucratic arms. The village would function as the limit to any industrialization or mechanization introduced by the state into the framework of social existence. While Gandhi urged that people should develop their spirituality, the political structure that he envisioned for India would preclude the possibility of the hegemonic imposition of any doctrine, religious or nonreligious, on Indian society.

In terms of the vocabulary of field and capital, Gandhi's perspective can be reframed as such: both the modern nation-state and science and technology sought to reduce the complex character of human sociality—embodied and organic, intellectual as well as material—to forms of capital that were accumulable and exchangeable. But for Gandhi, spirituality and morality, as the very core of individual and social human existence, were irreducible to capital and, indeed, were the very antithesis of capital. The spirituality and morality of the individual and community were anchored in and inseparable from a continuous and lifelong endeavor of reflective practice and ethical action. To reduce spirituality to a kind of capital, in the manner that the railways had done by enabling the easy accumulation of karma, would inevitably lead to the destruction of spirituality. This was Gandhi's precise objection to Western modernity, to the modern-nation state, and to science and technology: they had eroded the spiritual and moral bases of Indian civilization.

The social, cultural, and ecological economies of the local community, as understood by Gandhi, did not follow the paradigm of the market, which was an amoral and immoral system of exchange, production, accumulation, distribution, and profit. In *Hind Swaraj*, it is this marketization of all aspects of human life under colonial modernity and science and technology to which Gandhi strenuously objects. The marketization of ideas was enabled by the printing press, the railways were culpable for the marketization of religion, and the steam engine fostered the unequal accumulation of wealth and profiteering by allowing the transfer of grain during the time of famine. The marketization of the bubonic plague, roguery, and "evil" were, by implication, inevitable outcomes of the forces of Western modernity.

Finally, Gandhi's fear that people would lose independence of thought and action in allowing the state to think for them appears as a prescient anticipation of Bourdieu's argument that the citizens of the state, through their practices, replicate, legitimize, and socially entrench state categories of thought. The political structure of the decentralized Indian nation proposed by Gandhi would deny the possibility of the statist accumulation of metacapital. With the broader political structure constraining the one within it—for example, the district constraining the village—neither the state nor any community (or group of communities) would be able to impose its hegemonic vision on any another. Gandhi's convictions, however, rest on essentialized assumptions about human nature, the Indian temperament, and the Indian village. He assumed that the inhabitants of villages would inevitably reject science and technology and that villages were the authentic embodiment of the spiritual and nonmaterialistic Indian temperament. By the very nature of its being, village India, as Gandhi saw it, would offer a natural political alternative to the modern state. In contrast, for Veer Savarkar, a nationalist contemporary of Gandhi, the modernity of the independent Indian nation would be defined by an overarching, centralized ideology: the doctrine of Hindutva, or Hindu nationalism.

Savarkar: Modern India as a Military Hindu State

It is one of the great ironies of Indian history that the man credited with the founding of a nationalist ideology whose character is often understood as essentially religious was himself an atheist. The enigma of Savarkar can be deciphered to some extent by locating his thought in contrast to that of Gandhi. A revolutionary nationalist, Savarkar shared Gandhi's objective of Indian independence but strongly disagreed with Gandhi in his understanding of Indian society and on the question of the means by which to achieve national freedom. In contrast to Gandhi's insistence on nonviolent struggle, Savarkar advocated violence as a means of anticolonial resistance. In opposition to Gandhi's emphasis on Hindu-Muslim unity as a necessary precondition for independence and the viability of a free India, for Savarkar, real freedom for India would be achieved not only by driving out the British but also by curing Indian society of the effects of Islamic invasions. As scholars suggest, Savarkar's text, *Hindutva*, where he detailed his theory of the Hindu nation, may possibly have been written as a response to Gandhi's ideas.[18] Whatever its inspiration, Savarkar's text sought to establish the primacy of Hindu claims to India, not on the basis of Hinduism but on the basis of "Hindutva," which literally meant Hindu being or Hinduness.

As is well known, Savarkar defined Hindutva in broader terms than just Hinduism as religion. Savarkar argued that Hinduism was "only a derivative, a fraction, a part of Hinduism."[19] At once a theory of origin and authenticity, Hindutva as the totality and essence of Hindu social existence stood for an ethos, culture, and worldview—a way of life—with its roots in the Aryan civilization of ancient India.[20] "Hindutva," Savarkar held, "embraces all the departments of thought and activity of the whole Being of our Hindu race."[21] This historical and social reality did not, however, extend to all the inhabitants of India: the conditions for qualifying to be a member of this imagined community were "geographical origin, racial connection (rather ambiguously specified), a shared culture based on Sanskritic languages, and 'common laws and rites.'"[22] Those who fit these characteristics could belong to India, while those who did not, such as Muslims and Christians, were outsiders.[23] Even though some Indian social groups did not recognize or follow social institutions such as the caste system, this did not make them foreigners, because their blood was Hindu blood. As Savarkar argued,

> The Sanyasis, the Aryasamajis, the Sikhs and many others do not follow the system of the four castes and yet are they foreigners? God forbid! They are ours by blood, by race, by country, by God…We Hindus, are all one and a nation, because chiefly of our common blood—"Bharat Santati."[24]

But for others, such as Muslims and Christians, there was no way to qualify as Indian except by rejecting their own faith, even if Hindutva claimed to be a doctrine of cultural universalism and not one of narrow religious identity.

Indeed, despite Savarkar's claim that the ideology of Hindutva was not primarily a religious doctrine, religion very much remained part of the equation: one test of loyalty to the Indian nation was the acceptance of India by Indian Muslims and Christians as their *holy* land.[25] The title page of *Hindutva* proposes this conflation of religious, cultural, and national identity: "A HINDU means a person who regards this land of BHARATVARSHA, from the Indus to the Seas as his Father-Land as well as his Holy-Land that is the cradle land of his religion."[26] With their allegiance to alien faiths that had originated outside the cradle of Aryan civilization, Indian Muslims and Christians could never count as real Indians, that is, as culturally and nationally Hindu. But even an acknowledgment of India as their holy land would not enable Muslims or Christians to qualify as cultural-national Hindus. As Neufeldt argues, for Savarkar, one could become Hindu "through defense of Hindu culture, through intermarriage, or through adopting India as one's motherland or holyland. But there is one important qualification: To be Hindu one must subscribe to an indigenous dharma. A foreign dharma will simply not work."[27] The question of religious identity, while subsumed under the category of culture in Savarkar's elucidation of Hindutva, functioned in effect as the main principle of division between real Indians and non-Indians.

For Savarkar, the ideology of Hindutva would be at once the inspiration, method, and rationale of the independent Indian nation-state. Savarkar posited that despite the violence wreaked on their land by outsiders, especially by the Muslims and the British, the spirit of the Hindu people as a cultural community had endured, even if it had withstood severe body blows.[28] He argued that independent India had to be a strong nation-state, one that would finally allow the Hindu ethos to flourish unhindered; the destiny of the Hindu nation-in-the-making that had been interrupted over millennia by successive invaders would be realized through the apparatus of modernity. In formulating his vision, Savarkar found a ready model in the writings of Guiseppe Mazzini, the Italian cultural nationalist:

> an ideological framework and a political philosophy that combined cultural pride and national self-assertion with a modernist outlook and a vision of a strong culturally homogeneous nation embodied by a unitary state—the vision of the making of a modern Italian nation-state.[29]

Toward this end, Savarkar also strongly advocated compulsory military training for Indian students and the development of a military-industrial complex.[30]

The power of the nation-state and technology (embodied as the military-industrial complex) derived from their instrumental potentialities. Securing the Indian nation was not merely a question of ownership of territory: the cultural unity of the Hindu nation was the real issue.[31] The modern nation-state could be imagined in terms of a principle of cultural ownership, and territorial unity would simply be the spatial representation of this cultural unity, the empirical correlate of a self-evident historical truth that had survived as spirit or ethos.[32] The concept and structure of the modern nation-state allowed for the successful realization of the Hindu historical spirit. The military-industrial apparatus would keep the flame of the Hindu spirit alive in the independent Indian state. It would secure cultural unity through the inculcation of the qualities of aggression, cultural pride, and patriotism in the Hindu youth of the nation. Through the accumulation of force and techniques of self-defense and warfare, it would protect the Hindu nation from ever falling victim to the hands of cultural-religious invaders. This sentiment is expressed in Savarkar's two-fold cry, his "message to Hindudom on the eve of this 59th birthday of mine...Hinduise All Politics & Militarise Hindudom!!"[33] With the militarization of Hindudom, India would undergo a profound transformation, yet, paradoxically, the transition to modernity would enable the truest manifestation of her essence.

On the one hand, the modern nation-state, as understood by Savarkar, would not evolve through the accumulation of metacapital from exchange of capital by citizens in different fields. Rather, the state would simply be the manifestation of a fully developed cultural capital—Indianness or Hinduness—possessed by its legitimate citizens, that is, Hindus as defined by Savarkar, and deriving from time immemorial. The state, in other words, was envisioned fundamentally as a national-cultural field. For the religious minorities that lacked cultural capital, the conditions for obtaining cultural capital—conversion to Hinduness through the adoption of an indigenous faith—would be categorically spelled out even before the modern Hindu-Indian state would come into existence. The task for the state would be to implement the rules of this cultural field, which had an independent and coherent existence prior to and beyond the reality of the state.

On the other hand, one could also argue that Savarkar understood the potential and ability of the state to inculcate its citizens with certain categories of thought. The state was valuable for Savarkar because of its coercive reach and capability: the policies of the state—for instance, the development of the military-industrial complex and mandatory military training—would instill cultural pride in its citizens. As the powerful incarnation of the Hindu ethos, the Hindu nation-state would ensure the perpetuation of the ideology of Hindutva through inculcating the doctrine in all its citizens. Backed by the power of the state, the essential premise of Hindutva would become reality: minorities would either convert to Hinduism or would leave the nation. As the cultural field incarnate, the state would simply exist to optimize its own functioning. In Savarkar's vision, the role for science and technology, in the form of the military-industrial complex, is defined in the context of the cultural-national field. It is a means for protecting the existence of the field and enabling it to function smoothly.

Like Gandhi and Savarkar, Nehru, too, saw the state as a powerful agency—an actor with a monopoly over species of capital—with the ability to influence and transform the lives of its citizens. While his conception of an independent India visibly differed from that of Gandhi, on one basic issue he was in agreement with his mentor: Indian identity could not be defined in any narrow and exclusive religious, cultural, or ethnic terms. In contrast to Gandhi, however, Nehru saw the centralized nation-state and science and technology as positive agents for change and placed his unqualified faith in both these ideas.[34]

Nehru: A Secular, Socialist, and Scientific India

For Nehru, of a younger generation of nationalists than Gandhi and Savarkar, science itself would enable the transformation from the anticolonial to the postcolonial, even as it would allow the new nation to successfully express its unique ethos. Nehru stressed the quality of scientificity, or the "scientific method," which stood for a rational, critical temper and enlightened outlook toward the world that could help transform both the individual person and the larger society.[35] While appreciating the need for spirituality, Nehru was strongly critical of what he saw as the deleterious effects of religion on Indian society; organized religion in particular was the very antithesis of scientific values.[36] Yet Indian civilization was especially well-suited to developing the scientific outlook. While India had "a greater distance to travel" than the West to reach this goal, Nehru claimed that "there may be fewer major obstructions on our way, for the essential basis of Indian thought for ages past, though not its later manifestations, fits in with the scientific temper and approach."[37] Once science was introduced into Indian public life and consciousness, science and religion would harmonize with each other as perfectly compatible aspects of Indian culture. It was such a transformative scientific temper that needed to be adopted by the independent state and cultivated in the people of India in order to achieve a universal modernity. Technological development itself had to be anchored in rational and appropriate—and, in Nehru's sense of the term, scientific—philosophies of political and economic governance. Nehru's ideologies of choice in this regard were secularism and socialism.

A vital challenge for the new nation was to find a political framework that would be adequate to India's pluralistic character. Yet if India's strength flowed from her teeming diversity, the differences between her many peoples—most notably religious differences—also threatened to fragment the nation-state. Secularism, as opposed to a narrowly defined nationalism, appeared to Nehru to be the only viable political framework for India. Khilnani suggests that the genius of Nehru's conception of secularism lay in its radical inversion of any core hegemonic idea of what it meant to be Indian.[38] Nehruvian secularism proposed that "India was a society neither of liberal individuals nor of exclusive communities or nationalities but of interconnected differences."[39] This antidefinition of Indianness was based on Nehru's unshakeable conviction that "secularism was not a substitute civic religion, still less a political project to remoralize society by effacing religion and stamping a secular identity on all Indians."[40] The Nehruvian conception of secularism was not merely drawn on Western models but was distinct from them.[41] Indeed, Nehru's writings reveal an acute understanding of the dangers of a parochial nationalism and of the necessity that India should not traverse the path of other nations in this regard.[42]

Rent by interreligious strife, India was also plagued by problems of severe inequality. The problems of inequity to a large extent were derived from and maintained by caste hierarchies and feudal land relations and could neither be seen as purely economic matters nor purely be solved as such. The economic imperative, accordingly, had to be yoked to an imperative of social justice. Nehru's answer to the problem was a humane socialism. Here, too, Nehru understood socialism as consonant with the ethos of Indian social life. Indian social life, in Nehru's view, had always emphasized the collective good over individual advancement. The Indian consciousness had not been contaminated with the ideologies of individualism and the profit motive to the extent of the West.[43] The logic of "Democratic collectivism"—which was consonant with traditional Indian modes of sociality that privileged the group over the individual—controlled by the state would lead to "far more equitable sharing and a progressive tendency towards equalization."[44] Indians were, thus, already historically conditioned and culturally socialized into psychologically internalizing the value and practice of socialism. In contrast to Gandhi, Nehru maintained that the village could not be an autonomous economic entity in independent India. But socialist ideology would catalyze the stagnant village, barely held together by some dimly lingering memory of communal good, into a viable *political* unit.[45] It would bring rural India in line with the vision of the state while restoring the traditional communal vitality of the village.

In the Indian state imagined by Nehru, scientificity, as a worldview, would operate as the nomos or the founding principle of vision of the state. The independent Indian nation-state would stand as an incarnation of this nomos. Scientificity, relatedly, would be the metacapital, held in monopoly by the state, with the force to dictate and shape the nature and value of other kinds of capital: economic, cultural, educational, social, or political. The state would seek to transform all aspects of social existence—every field—according to the principle of scientificity, enabling accumulation as well as exchange. It would promote an enlightened temper among its citizens through commitment to the rational doctrines of secularism, progress, and development; the state-subsidized educational system would, likewise, stress scientific and technological education; the state would usher the Indian masses, urban and rural, into modernity through instruments and technologies of industrialization and development planning; and the centralized socialist state would accumulate economic capital and redistribute it on an egalitarian basis.

In concluding this section, a few observations bear noting. For all their disagreements about the role of science and technology, it may be reductive to view the respective thought of Gandhi and Nehru as categorically opposed. They shared core beliefs about the nature of Indian civilization as imbued with a unique ethos: a set of distinctly Indian attributes of social plurality and coexistence that would give the project of Indian nationhood its substance and character. As Prakash points out, "there is no denying the difference between the two men. Undoubtedly, Nehru set great faith in the power of modern science and technology, while Gandhi did not."[46] However, both Gandhi and Nehru articulated their nationalism as a "critique of Western modernity."[47] And both "contended that India's modern nationhood would be the fulfillment of its timeless unity as a community; India would be at once modern and not modern, at once represented and not represented in a state."[48] Similarly, while Savarkar's theorization of Hindu nationalism is often criticized as an uninspired derivation of European cultural nationalism, in one respect his thought inhabits the same discursive field as Gandhi and Nehru. He also located the future of the Indian nation in a distinctly Indian character, although this was derived from a Hindu rather than pluralistic ethos. In Savarkar's vision, too, the cultural imaginary called India, the soil that nourished the nation-state, would always exceed the latter as a cultural-historical surplus beyond its concrete representation in the institutions of modernity.[49]

As we know, Nehru's vision of the modern Indian-state triumphed over the alternative conceptions of Gandhi and Savarkar. But Gandhi's fears about the centralized nation-state and science and technology have proven to be prescient in many ways. And despite the marginalization of the Hindu nationalist agenda from the political mainstream in the decades after independence—in no mean measure because of the association of Gandhi's assassin, Nathuram Godse, with organizations dedicated to the ideology of Hindutva—Hindu nationalism would stage a resurgence in the late 1980s and 1990s. Savarkar's propositions of a militarily strong Hindu India would find endorsement in the cheers that greeted the tests announcing India's nuclear bomb, which were conducted in 1998 under the rule of the Hindu nationalist BJP-led National Democratic Alliance.

The Technological Field and Capital in Independent India

This section describes how the policies of the Nehruvian state have shaped a technological field in postcolonial India and how the technological field is located with regard to the economic, cultural, and educational fields. I focus on three foundational aspects of the nomos of the technological field as it has been shaped in postcolonial India by state policies as well by the social reception and impact of these policies. The first aspect is the fact that participation in the nationalist project of scientific and technological development was rewarded by the Indian state with the gift of a special kind of cultural capital. The second aspect is the state's emphasis on supporting higher education in the medium of English as opposed to supporting primary education. In independent India, fluency in English has been a prerequisite for securing both a scientific and technological education and occupation. And the third foundational aspect is the perpetuation of the traditional value attached to an overseas education by the middle and upper-middle socioeconomic segments of Indian society, a legacy of colonial-era attitudes of Indian elites toward education. These characteristics of the nomos of the technological field can be understood as the terms of entry and participation for the accumulation for capital, the precondition or principle of vision on which the field is grounded.

Following independence, the modern Indian nation-state was conceptualized as a centralized agency that would promote national development through a three-point agenda of socialism, secularism, and scientific and technological development. Industrialization, economic growth, and the emphasis on scientific and technological research would help India gain parity with the advanced, developed nations of the West. Technology would help the socialist Indian economy become self-sufficient, and within a few decades, India would no longer have to rely on importing food for its masses. It was believed that economic self-sufficiency would also secure a politically independent voice in the global community for India, a poor, developing nation likely to be pressured by powerful nations such as the United States. The socialist-secular political framework would enable all Indian groups to benefit equally from the rewards of industrial development. Not only would state-owned industries provide large-scale employment, but economic subsidies, subsidization of higher education, price controls, salary ceilings, and taxation structures would ensure an equitable distribution of wealth. It was also hoped that those educated in India's premier technological institutions would work for the well-being of the Indian masses. An accompanying belief was that the rewards of a socialist system would help establish a socialist sensibility amongst the Indian people. As a result, traditional ills such as caste inequalities and oppression would wither away along with reductions in widespread poverty and economic disparities. Nehruvian secularism would provide a framework for establishing equality among India's different religious communities. Socialism would support the secular character of the Indian state by helping create a more just society. And science and technology would be critically important instruments for realizing this fervidly dreamed-for Indianness.

With the onset of independence, the Indian state implemented a program of centralized planning and state ownership of heavy industry. Dams, steel plants, hydroelectric plants, and factories constructed across the country were the new symbols of modern India. Industrial development was given pride of place over agricultural development.[50] Over fifty years since independence, the Indian state also set up numerous institutes for industrial training, higher education in engineering and technology, and forty-eight national laboratories for applied scientific research.[51] As Nandy argues, in independent India, science was authorized as a "reason of state."[52] It would not be an exaggeration to say that science was conferred with a quasi-religious sanctity. Visvanathan notes that "Dams and laboratories were literally the temples of India."[53] The policies of the Nehruvian state also ensured that the institutions of scientific research and development planning were placed outside the political process and immunized from the demands of popular politics.[54] A scientific and technological education, especially from the premier educational institutions, operated as a marker of talent and capability, ensured job opportunities, and promised lifelong security. Those holding an education or an occupation designated as scientific or technological were perceived as working in the service of the nation.

Scientific or technological educational qualifications and occupations directly guaranteed educational, cultural, and economic capital. A defining characteristic of the nomos of the technological field was articulated at this point: the very possession of technological capital was positively associated with service to the nation or patriotism. The framing of the Nehruvian project in accordance with socialist goals meant that science and technology, by association, were authorized as instruments of social justice. Institutions and individuals working in science and technology in independent India were defined as the agents of a progress that could be measured in empirical terms. Importantly, they were also validated as moral agents, embodying a rationality of perspective and nobility of purpose toward human society at large. The nomos of the technological field thus linked science and technology both with the national good and a more foundational philosophical idea of the good or noble.

Nehru's definition of scientificity as a reflection of state-approved philosophical and practical ability was enforced by state policy as the principle of exchange between the technological field and the economic, educational, and cultural fields, respectively. The technological field made thinkable a set of relations between the economic, the educational, and the cultural domains of Indian social life; it was the common thread that tied these fields together. Science (with technology) was not just "a reason of state," in Nandy's memorable phrase, but it was also a principle of ordering and refashioning society and culture. Technological capital in Nehruvian India was of paramount significance; as state-sanctified metacapital, it was currency that could be simultaneously cashed in as different kinds of value. Exchanging it as educational and economic capital did not remove it from the play of cultural capital. The more technological capital was exchanged by its holder, the more its value multiplied. An education in one of the well-known national institutions of science and technology, followed by a prestigious job in technology, would only increase the cultural capital of the holder.[55]

A second fundamental way in which the policies of the Indian state have impacted the nomos of the technological field since independence is through a corresponding articulation of the nomos of the educational field—specifically, through the privileging of higher education over primary schooling and the privileging of education in the medium of English. This has meant a privileging of the interests of certain elite segments of Indian society who share and, in turn, endorse the vision of the state. As Bourdieu has demonstrated through his research on the French educational system, it is through the educational field that existing social hierarchies are accepted as the natural order of things and social inequality is reproduced as well as granted cultural legitimacy.[56]

Shurmer-Smith points out that India spends a minuscule amount on primary education as compared to its heavy investment in higher education.[57] Many of India's quality institutions of higher education, whether in engineering, medicine, the sciences, or the humanities, are either government institutes or are financially supported by the central or state government through grants. College education in a government institution is highly subsidized, and the cost is negligible compared to that of an education at a private institution, whether at the school or college level. Since English is the medium of instruction in premier educational institutes, and since English has been the language of commerce in independent India, fluency in English is an essential prerequisite for obtaining a quality higher education in India. Shurmer-Smith points out that the Indian education system has privileged those who can afford a private schooling in English, the "affluent middle classes and above," who spend considerable sums of money toward providing such an education for their children.[58] Fluency in English combined with a private education provides educational capital for specialized learning and training. Capital obtained from higher education directly translates into economic capital.

At the same time, fluency in English has also translated directly into cultural capital in independent India—a reflection of British colonial rule and, equally, of the fact that Indian anticolonial nationalist consciousness in its inception reflected the ideologies of educated, middle-class, and English-speaking urban elites. An individual's ease with the English language, especially the ability to converse with the "right" accent, is a signpost that one hails from an Indian cultural elite. This, in turn, might suggest that one's family has for several generations belonged to a privileged segment of Indian society—for example, those who have had access to a Western-style education. The independent Indian state, through its support of educational instruction in the medium of English, has reinforced such perceptions.

The emphasis on scientific and technological education exemplified in the policies and rhetoric of the Indian state can be viewed as the "objective" dimension of the educational field in India. This "objective" logic of the value of a scientific or technological education has been incorporated by Indian society in its attitudes to scientific and technological education in the postindependence era. A number of indicators bear testimony to this: the intense competition for seats in engineering and medical colleges; the steady mushrooming of private engineering and medical colleges with facilities and infrastructure of less than ideal quality; and the willingness of the lower-middle, middle, and upper-middle socioeconomic segments of Indian society to spend large sums of money on private tuitions to prepare their children for entrance examinations to these institutions.[59] Data provided by Aggarwal show that the category of "[e]nrolment in science courses as a percentage of total enrolment" in university education grew from 25.7 percent in 1960–1961 (0.19 million of 0.74 million) to 44.7 percent in 1973–1974 (1.19 million of 2.67 million).[60] And yet, it is not the case that the orientation of various segments of Indian society toward the educational field has been wholly determined by the "objective" dictates of state policy. For if scientific or technological educational qualifications have consistently continued to be prized in Indian society, it is equally true that those who obtained such an education have not necessarily put it to the use of national progress. Especially since the 1970s, large numbers of Indian graduates—whose education has been subsidized by the state—have chosen to migrate overseas, mostly to the United States, for the purpose of further educational qualifications and careers. As an article about India's premier technological institutes notes, "critics have carped about the Indian taxpayer subsidising the education of IITians only to see them take the first flight out to the US of A."[61] While an education from an Indian Institute of Technology continues to operate as a signifier of social status in postcolonial India, an IIT degree leading to a job in the United States is considered even more prestigious. In this manner, the structure of the educational field in postindependence India has been shaped as much by the subjective aspirations and motivations of social groups as by objective forces. Indeed, given that scientific and technological qualifications continue to be highly valued in India even today, there seems to be an "adjustment" between objective and subjective forces, seemingly resulting in a kind of equilibrium that gives the educational field in India its particular character. The subjective aspiration of educated individuals to migrate to the United States has itself become a kind of objective determinant of the field.

The desire of Indians for an overseas education and occupation is also itself the product of a cultural history. This is the third defining characteristic of the nomos of the technological field: the value that Indian elites have traditionally attached to an overseas education. As Shurmer-Smith argues, from the colonial period onwards,

> the upper levels of metropolitan Indian society have been conscious of the importance of operating on an international scale as a means of preserving their impermeability to the classes below. Education has long been an important aspect of this and during the British Raj many upper-class families strove to educate their sons expensively in the most exclusive British boarding schools and universities.[62]

While more recently this trend has shifted toward the United States, large numbers of students continue to go abroad for higher studies.[63] An overseas education brings with it educational capital and economic capital, since such qualifications "put a candidate in a very strong position when competing for senior positions in national and international corporations in India."[64] But the value of an overseas education seems, in good measure, to be cultural. Historically recognized as a marker of privilege and of membership of an economic and cultural elite, an overseas education is a guarantor of much-sought-after "prestige."[65] Remarking on her conversation with a few hundred youth at education fairs in India, Shurmer-Smith notes that for these Indians, "the overseas education was desirable *because* it was expensive and because it was overseas."[66]

In independent India, an overseas education promises more cultural capital than an education from even a top-quality Indian institute.

These complexities may partially explain the phenomenon of "brain drain," the migration, since the 1970s, of large numbers of qualified Indians from the nation's premier educational institutes to the West for the purpose of higher studies and pursuing careers abroad. Khandelwal points out that "in the 1970s, the Immigration and Naturalization Service was recording an astonishing and almost unprecedented rate of about 80 to 90 percent of working Indians as professionals."[67] Khandelwal points out that since the 1970s, the Indian immigrant population in the United States has considerably diversified in its composition, yet the majority are still professionals or students seeking professional qualifications. The migration of such Indians might be propelled by their desire to add cultural capital to their pool of educational and economic capital through using their Indian qualifications to acquire an additional qualification overseas. The accumulation of different kinds of capital enabled by migration to the United States, in turn, reaffirms the value of Indian scientific and technological qualifications.

A pattern can be distilled from these arguments, which allows one to grasp the character of the nomos of the technological fields in independent India and, relatedly through the matrix of exchange, that of the economic, educational, and cultural fields as well. The structure of each field has tended to privilege economic and cultural elites, specifically those who have been able to afford private schooling in the medium of English and possess a scientific and technological education. Yet this privilege has been affirmed by the state in the name of the nation, since both scientific and technological education and fluency in English have been authorized as (a) instruments in the service national project of material and moral progress and (b) as qualities of universal value. The clamor among Indians for an overseas education may seem to contradict the Nehruvian imperative of national self-sufficiency and progress. The state endorsement of education in the medium of the English language also betrays an ambivalence on the part of the state with regard to its project of encouraging the Indian population to develop the mysterious quality of the rational, enlightened temper or "scientificity." On the one hand, the Indian nation-state proposes that those segments of the Indian society who are already privileged, economically and culturally, are best suited and positioned through the benefit of education to promote the virtues of enlightenment and scientificity among the poor, uneducated, non-English-speaking masses. On the other hand, as is not too difficult to see, this axiom directly contravenes the professed socialism of the Indian state. But these contradictions do not impair the functioning of the technological, cultural, economic, or educational fields in independent India. For, as Bourdieu has so convincingly demonstrated, the nomos of the field remains outside the ken of the thinkable and questionable even as it structures the limits of thought and action within the field.[68] For the field to function, the illusion must be maintained that the essential premise for the exchange of capital is fair. In the Nehruvian state, those already possessing economic and cultural capital were authorized as the heirs of technological capital. Indeed, it was assumed that the English-speaking elites were already endowed with the capacity for a scientificity deemed synonymous with a rational, cultivated, and enlightened outlook. This points to an implicit naturalization of the structural inequities that characterized fields—seen, for instance, in a discourse of meritocracy routinely invoked by Indian elites to justify their privileged status—even if these inequities were also acknowledged as historically constituted. In postindependence India, in the ultimate analysis, it was technicity—or the accumulation of specific professional technological skills—that counted as technological capital.[69] One could not, in other words, claim an entitlement to technological capital on the grounds of possessing a rational "scientific" outlook, without scientific or technological qualifications. At the same time, technicity itself was garbed as "scientificity," with its associations of service to the nation. In other words, it was, paradoxically, the discourse of scientificity that allowed the principle of technicity to be established as the logic of accumulation of technological capital.

Although the character of the technological field has changed significantly since its origins in the colonial era, the nomos of the field continues to be characterized by a majoritarian discourse as it was at its founding moment. What is common to both the discourses is the privileged access of certain already-privileged groups to the realm of technology and the validation of their privilege in terms of an idea of service to the nation. In the decades since independence, the Indian nation-state has initiated and effected transformations in the technological and other fields, but, importantly, has also preserved continuities with the nomos of these fields. The description of characteristics of the nomos of the technological field already suggests a profile of the class of Indians that has been optimally positioned to accumulate technological capital. This chapter concludes with a more detailed description of the qualities of this class.

Mapping the Technological Class in Independent India

We can already identify some attributes of the theoretical class that is structurally privileged with regard to the Indian technological field: it corresponds to urban, metropolitan, cultural and economic elites, the English-speaking middle, upper-middle and upper-socioeconomic segments of Indian society that have been the biggest beneficiaries of the policies of the independent Indian state. Following Bourdieu, it is important to remember that this is not a real, empirical class that is numerically distinct and mutually exclusive from other social classes. A person possessing economic capital, for instance, might lack the necessary cultural capital—fluency in English—to obtain a scientific education. Correspondingly, a person possessing cultural capital may lack the economic capital to obtain a private schooling that would enable him or her to pursue a specialized higher level degree in science and technology. One can sketch the theoretical class of Indians possessing technological capital (or the capacity to accumulate capital) with approximate reference to some information about India's elites, in the context of India's population of over 1 billion people.[70]

Given the unreliability and absence of indicators, such as tax, income, and consumption statistics, there are methodological problems in estimating the size of India's "middle classes."[71] Vanaik estimates that India's socioeconomic elites can broadly be defined as the top 25 percent of income-earning segments of Indian society that, according to different parameters, number anywhere between 100 and 250 million.[72] Harris-White points to data from the 1991 census that indicate that "fewer than 12 per cent of the population lived in metropolitan cities, the headquarters of corporate capital, the 'habitus' of the globalised, economic, technological and political elites."[73] By the "metropolitan" population, Harris-White is referring to a subset of the urban population, specifically the populace that lives in cities with more than 200,000 inhabitants.[74] Along with this information, census categories and the criterion of fluency in English provide some rough parameters for conceptually dividing the technological class into three broad fractions.[75] The core consists of a metropolitan, urban, English-speaking elite population, which would possess the greatest access to technological capital via possession of economic and cultural capital. This is followed by a secondary tier that is also an urban metropolitan elite but not as proficient in English as the inhabitants of the first tier. The inhabitants of the secondary tier possess economic capital but not the same level of cultural capital as the first tier. The second tier is followed by a third tier that possesses minimal economic capital through its location in an urban but nonmetropolitan, nonslum, economy. Both the second and third tiers stand at a serious disadvantage in comparison with the core group because of an absence of cultural capital stemming from a lack of fluency in the English language. This distribution of access to technological capital and capacity to accumulate and exchange technological capital are at once a differential distribution of privilege.

Marxist scholarship offers some interesting theoretical conceptualizations about the composition of Indian elites since independence that are relevant for our purposes here. Bardhan argues that the independent Indian state through its policies has created dominant elite classes.[76] Challenging the orthodox Marxist perspective that the state is "a direct tool of the capitalist class," Bardhan focuses his analysis on the conditions and mechanics of accumulation of wealth and not on the state policies and patterns of distribution of wealth.[77] Bardhan argues that there are three kinds of groups that have benefited, in terms of wealth accumulation, from the policies of the Indian state: urban industrial elites, agrarian elites or rich farmers, and bureaucratic-professional elites comprised of white-collar workers and those occupying important positions in the Indian state bureaucracy. He suggests a tendency on the part of the Indian state to side in its policies with the urban industrial and bureaucratic elites against the rural elites.

In economic terms, we can, therefore, define an urban professional class as a distinct subset of India's elites and one that is privileged by the state. Khilnani describes the rise of a new class of Indians since the 1970s, whose achievements and aspirations are symbolized by the city of Bangalore.[78] He states: "Since the 1970s," Bangalore

> has experienced rapid growth and a new middle and upper middle class has emerged. This is based not on the traditional sources of wealth in independent India—control of land, bureaucratic office or industry—but on *professional and technical skills*.[79]

At the same time, India's urban "professional" elites can also be conceptualized as encompassing communities specializing in traditional and networked economies of services and trade. The conflict of interests among Indian urban elites are not necessarily only a matter of economic superiority but may also reflect differential allocations of cultural capital based on traditional social hierarchies of caste.

Indeed, the caste and socioeconomic class combinations that have comprised the numerically sizeable Indian elites since the colonial era and that have collectively been "the prime beneficiaries of India's post-colonial political economy" are quite complex.[80] These beneficiaries can be divided into two categories. The dominant segment consists of "the Brahmins, Rajputs, Kayasthas, Khatris etc. whose power, in rural areas, originated from control of land, and in the towns and cities, from the benefit of early English education and employment in the colonial administration."[81] This segment, "who in urban areas tend to occupy the middle and upper levels of class hierarchy, came to dominate government, bureaucracy and the professions after independence."[82] This highly professionalized segment of Indian society can be understood as both upper caste and upper class. In the broadly defined category of elites also falls a secondary group, "the somewhat lower-ranking and highly urbanized castes of merchants, moneylenders, shopkeepers and small-scale entrepreneurs" who specialize in commerce, traditional occupations of finance, and trade.[83] This mercantile strata of the elites, while still upper caste, is of relatively lower caste status than the professional strata, and, in economic terms, is "mostly lower middle-class."[84]

Against this backdrop, we can now visualize a composite picture of the Indian technological class as a theoretical entity. The inhabitants of the technological class occupy different positions in the cultural field, both in terms of fluency of English and caste identity. Nevertheless, they need to possess a basic amount of cultural capital—broadly upper-caste identity and some familiarity with English—to enter the technological field. While India's rural elites also possess substantial economic capital that might enable entry into the technological field, in the view of the state, it is the interests of the urban professional classes and urban industrialist classes that are of greater importance. Between the urban industrial classes and the urban professional classes, the possession of formal educational qualifications is arguably of greater import to the latter, given that they make their livelihood through salaried employment as opposed to entrepreneurship or business. The upper-caste and upper-class professional segment also is literally vested in the state through monopolizing positions in government and the state bureaucracy. This possibly explains the strong resistance of this elite group to any government initiative that seeks to increase through affirmative action the percentage of seats and jobs reserved in state educational institutions and the state-owned employment sector for historically disadvantaged segments of Indian society. Juxtaposing the arguments of Bose and Bardhan, we get a sense of how the Indian state has refashioned the relationship between cultural capital and economic capital through the forging of a technological field. Caste hierarchies as a principle of social organization are not a creation of the policies of the Indian state, but the Indian professional elites—who are also an elite group in terms of caste—emerge, ironically, in the context of the socialist policies of the state. Even as the state officially seeks to eradicate casteism from Indian society through technocratic development, caste, cultural, and economic privilege transfer over to the technological field as significant quanta of capital, tantamount to a claim of ownership. The professional and technocratic elites, one might argue, manage to inculcate in the state their categories of thought.

Yet the definitions of cultural capital articulated by the elites in their practices, and tacitly affirmed by the Indian state, have not been uncontested in independent India. As we will see, while the privilege accorded to technology and technological capital in India does not diminish in the 1990s, other Indian groups, within and outside an elite, seek to redefine the very nomos of the cultural field and the economic field. In the cultural field, Indian secularism is progressively weakened by the rise of Hindu nationalism, and in the economic field, the socialism of the Nehruvian state is officially abandoned for a neoliberal reinvention of the Indian economy and society. The technological field and technological capital, however, continue to play a critical role in relation to both fields and in the imagination of the Indian state.

Endnotes

1. The Gandhian and Nehruvian visions of political modernity, the Indian nation-state, or development are often compared or juxtaposed against each other. Including Savarkar's idea of India within this evaluative framework reveals the extent to which the three models may share the same discursive field and also helps identify important points of divergence between them. Such an exercise requires suspending, temporarily at least, the belief that the Hindu nationalist project is the antithesis of the Nehruvian secular vision or that the understanding of Indian culture in each view shares nothing with the others.
2. The year 1991 can be taken as the moment when the Indian state officially embraces a neoliberal paradigm and turns away from the socialist framework of economic and political governance.
3. Gandhi's critique of modernity and technology has been extensively addressed by arguably his most insightful reader, the Indian psychologist and cultural critic Ashis Nandy. See, for example, Ashis Nandy, "From Outside the Imperium: Gandhi's Cultural Critique of the West," in *Traditions, Tyranny and Utopias: Essays in the Politics of Awareness* (Delhi: Oxford University Press, 1987), 127–162. Nandy argues that Gandhi "rejected technicism, not technology," that is, an instrumentalist understanding and use of technology whose *telos* is necessarily self-serving (137–141). Prakash argues that Gandhi's idea of the village as the unit of Indian society was informed by an appreciation of some of the benefits of Western technology. Gandhi envisioned Indian villages as free of disease, thanks to Western medicine, and even linked with rail and telegraph networks. See Gyan Prakash, "The Urban Turn," in *SARAI Reader '02: The Cities of Everyday Life*, ed. Ravi S. Vasudevan et al. (New Delhi: Sarai Media Collective, 2002), 3–4. See also his reading of Gandhi's critique of modern and industrial civilization in *Another Reason*, 214–226. In a fascinating essay, Visvanathan imagines what Gandhi's creative response to contemporary science might be. See chapter 7, "Reinventing Gandhi," in Visvanathan, *A Carnival for Science*, 212–244. Sahasrabudhey offers an interesting interpretation of *Hind Swaraj* as a critique of a particular mode of scientific theorization. See Sunil Sahasrabudhey, "*Hind Swaraj* and the Science Question," in *Hind Swaraj: A Fresh Look*, ed. Nageshwar Prasad (New Delhi: Gandhi Peace Foundation, 1985), 99–105.

The apparent tension in Gandhi's views about technology can be explained by the fact that he spoke of science and technology either as synonymous with a particular Western or colonial history *or* as a broader set of constructs that also provide the space for alternate trajectories. The same dual use of terms, arguably, is also seen in the way Gandhi conceptualized the modern and modernity.

4. M. K. Gandhi, *Hind Swaraj and Other Writings*, ed., Anthony J. Parel (Cambridge, U.K.: Cambridge University Press, 1999).
5. Across his vast body of work, Gandhi did not, however, explicitly articulate a systematic theory of the state. As Parekh points out, "Gandhi's theory of the state was uneven, in parts refreshingly original and profound, in others ahistorical and naïve." Bhikhu Parekh, *Gandhi's Political Philosophy: A Critical Examination* (London: Macmillan Press, 1989), 110. *See esp.* chap. 5, "Theory of the State," 110–141.
6. Ibid., 113. Parekh points out that by the 1930s, Gandhi's thinking about the state had somewhat changed. Gandhi's expectations that Indian communities would reform themselves towards the objectives of social justice had been belied by the failure of orthodox Hindus to implement his suggestions about caste reforms to abolish untouchability. Gandhi conceded that the state could possibly play a role as a moral actor in promoting socially just causes where hegemonic community ideologies might prohibit self-reform. Yet he retained his fundamental skepticism about the state, especially with regard to the state's monopoly over power and its coercive hold over its inhabitants (117–120).
7. Parekh states, "As he sometimes put it, the modern state was organised along 'scientific' lines and had reduced the government to a machine. Gandhi's criticism of it was an integral part of his larger critique of modern science" (111).
8. See chapter 1, "Critique of Modern Civilisation," in Parekh, *Gandhi's Political Philosophy*, 11–35, esp., 21–27. Gandhi's rejection of the Western manifestations of science and technology in *Hind Swaraj* also needs to be situated in terms of his analysis of the philosophical roots and instruments of colonial domination.
9. Ashis Nandy, *The Savage Freud and Other Essays on Possible and Retrievable Selves*, in *Return from Exile* (New Delhi: Oxford University Press, 2004), 181.
10. Human limits and human control are deeply related in Gandhi's understanding of humanness. In exceeding, or aspiring to exceed, their natural limits, humans risk losing their very humanness. Nandy observes that

"[a]s a general rule, Gandhi was against technologies which replaced the uniquely human aspects of man" (*Traditions, Tyranny and Utopias*, 138). Prakash notes that "the theme of the loss of human control to technology was Gandhi's constant refrain" (*Another Reason*, 217).

11. Gandhi, *Hind Swaraj*, 51.
12. Ibid.
13. Ibid.
14. Ibid., 47.
15. Sahasrabudhey, "*Hind Swaraj* and the Science Question," 102.
16. Gandhi, *Hind Swaraj*, 36–37.
17. Parekh, *Gandhi's Political Philosophy*, 136. Parekh identifies four main principles of Gandhi's articulation of the "spiritual economy," which are as follows:

    First, every man had a right to work…Second, economic life should be subordinated to and regulated by man's moral and spiritual needs… Third, since the village communities were to form the basis of the Indian economy, the nature, pace, and scale of industrialisation was to be determined by and subordinated to their requirements…Fourth, the means of production of the basic necessities of life should be collectively owned. (136–137)

18. Vinayak Damodar Savarkar, *Hindutva: Who Is a Hindu?* 5th ed. (Bombay: Veer Savarkar Prakashan, 1969). The first edition of the text was published in 1923. Neufeldt observes, "Indeed, Savarkar's biographer, Keer suggests, that *Hindutva* was, at least in part, an intense reaction to Gandhiism, particularly with respect to the treatment of the antinational demands of the Muslims." Neufeldt, "The Hindu Mahasabha and Gandhi," 139. Gandhi and Savarkar are known to have met at least once in 1909 (Puri, "Indian Nationalism and the Hindu Argument," 154). Puri suggests that Gandhi and Savarkar may have also met earlier in 1906 in London, "although there is no historical evidence for the meeting" (ibid., 152). See also Dhananjay Keer, *Veer Savarkar*, 3rd ed. (Bombay: Popular Prakashan, 1988). The first edition of the text was published in 1950.
19. Savarkar, *Hindutva*, 3.
20. For Savarkar, Hindu consciousness was "as old as the civilization that was fashioned in India by the ancient Aryans" and "therefore, at least as old as the *Rg Veda*" (Neufeldt, "The Hindu Mahasabha and Gandhi," 136).
21. Savarkar, *Hindutva*, 4.
22. Khilnani, *The Idea of India*, 161.

23. Ibid.
24. Savarkar, *Hindutva*, 39.
25. Vanaik defines Hindutva as a belief in a particular chain of reasoning—a discourse based on links between the four elements of ethos, religion, culture, and nation. "Ethos/spirit is at the heart of Hinduism, which is at the heart of, or coterminous with, Hindu culture (which defines Indian culture), which is at the heart of the Hindu nation (which defines the Indian nation)." Vanaik, *The Furies of Indian Communalism*, 151. Hardgrave and Kochanek define the doctrine of Hindutva in the following manner: "all Indians—including Muslims—are part of a Hindu nation and the gods and heroes of Hindu mythology are part of their patrimony." Robert Hardgrave and Stanley Kochanek, *India: Government and Politics in a Developing Nation* (Fort Worth, Texas: Harcourt Brace, 1993), 181–182.
26. Savarkar, *Hindutva*, title page.
27. Neufeldt, "The Hindu Mahasabha and Gandhi," 138.
28. In *Hindu-Pad-Padshahi*, Savarkar details a history of Maratha resistance and valor against the depredations of Islam and Muslim rulers as well as the English. The Maratha "achievement," Savarkar notes, "is bound to reflect its greatness on the history of Hindudom as a whole." Vinayak Damodar Savarkar, *Hindu-Pad-Padshahi or a Review of the Hindu Empire of Maharashtra*, 4th ed. (New Delhi: Bharatiya Sahitya Sadan, 1971), 18. The first edition was published in 1925.

    In Hindu nationalist discourse, Indian Jews and Parsis are described as guests who do not pose the same threat to the Indian nation as Muslims and Christians, although none of these communities fulfill the Hindutva criteria of Indianness. The Hindu nationalist M. S. Golwalkar, in his text *We or Our Nationhood Defined*, described Jews and Parsis as persecuted communities sheltered by Hindus and Hindustan. See M. S. Golwalkar, *We or Our Nationhood Defined*, 3rd ed. (Nagpur, India: Bharat Prakashan, 1945), 23–24. The first edition of the text was published in 1939.
29. Hansen, *The Saffron Wave*, 77. See also Khilnani, *The Idea of India*, 160.
30. Neufeldt, "The Hindu Mahasabha and Gandhi," 140, 142. Savarkar supported India's participation in World War I on the grounds that this would provide Indian youth with much needed military training and experience (ibid., 140).

> The call for militarization and the building of a military-industrial complex became an integral part of the Hindu Mahasabha platform from 1939 to 1941. One finds repeated references to the need for militarization, military training of the youth, and supporting a military-industrial complex in the resolutions of the Mahasabha. (ibid., 140)

31. Neufeldt points to the fact that "Savarkar saw the principle of territorial unity as deeply flawed" (ibid., 144). Savarkar saw the idea as derived from Western education and, in his opinion, Muslims were not committed to the principle either (ibid., 144).
32. Vanaik notes that the central feature of Hindutva is a "unity of culture and territory" (*The Furies of Communalism*, 150).
33. Vinayak Damodar Savarkar, *Historical Statements—V. D. Savarkar*, ed. S.S. Savarkar and G. M. Joshi (Bombay: Popular Prakashan, 1967), 1.
34. See Prakash, chapter 7, "A Different Modernity," in *Another Reason*, 201–226 for a nuanced analysis of the similarities and differences in the thinking of Gandhi and Nehru on these questions. As Prakash suggests, both the differences and similarities need to be located in the context of the problematic of articulating a coherent vision of postcolonial modernity.
35. In *The Discovery of India*, written in 1944, Nehru observes, "The scientific temper and approach are, or should be, a way of life, a process of thinking, a method of acting and associating with our fellowmen" (Nehru, *The Discovery of India*, 512). According to Nehru, science showed the way to freedom: "The scientific temper points out the way along which man should travel. It is the temper of a free man" (ibid.). Prakash notes that Nehru

    > was, as is well known, an avid advocate of what he called the scientific outlook...A much more expansive view of science than its conception as theoretical discoveries, Nehru's view projected science as a method that could be applied just as easily and effectively to order politics and economics as to understand nature. (*Another Reason*, 196)

    The scientific approach to social existence could explain and transform politics, society, and "life itself" (ibid.). Brown observes that

    > [a] sense of embracing the modern world was an essential aspect of Nehru's idea of India...It meant being prepared to embrace change, after reflection and rational discussion, and determining to take charge and use it for the good of the whole, in the fields of politics, society and the economy. Such a stance was, in his view, rooted in a basically scientific attitude to life. (Judith M. Brown, *Nehru: A Political Life* [New Haven, CT: Yale University Press, 2003], 189)

    Nehru's faith in science and insistence on the scientific temper has received fairly extensive attention from scholars. My focus is on critical points of comparison, similarities as well as differences, between his thought when compared to Gandhi and Savarkar's respective views on the subject.
36. Nehru, *The Discovery of India*, 513.

37. Ibid., 515. See also Prakash, *Another Reason*, 210–211.
38. Khilnani, *The Idea of India*, 171–172.
39. Ibid.
40. Ibid., 177–178.
41. Ibid., 167. The viability of Nehruvian secularism as a framework of governance for the Indian state and its value as a nationalist ideology are much discussed and contested topics in Indian scholarship. While Khilnani views Nehru in a sympathetic light, Nehru's idea of India has provoked a significant amount of criticism as well. My focus here is on Nehru's vision of India, as he saw it, which Khilnani sums up effectively. Two excellent collections on Indian secularism, which reflect the shifting contours of the legacy of Nehruvian secularism over a decade, are Rajeev Bhargava, ed., *Secularism and Its Critics* (New Delhi: Oxford University Press, 1998) and Anuradha Dingwaney Needham and Rajeswari Sunder Rajan, eds., *The Crisis of Secularism in India*, (Durham, NC: Duke University Press, 2007).
42. For example, in a 1928 speech to the All-Bengal Students Conference in Calcutta, Nehru cautioned that "the reaction against foreign rule" ran the risk of making Indians "narrowly" nationalistic. "And I have often wondered," Nehru said,

    > if there is any country in the world, any people, who do not fancy that they have a special mission for the world. England has her white man's burden which she insists on carrying, in spite of the ungrateful people who object and rebel; France has her mission of Civilisation; America is God's own country; Germany has Kultur; Italy has her new gospel of fascism; and Russia has her communism. (Jawaharlal Nehru, "From Address to Youth, All-Bengal Students Conference—Calcutta, September 22," in *Nehru: The First Sixty Years*, sel. and ed. Dorothy Norman [London: The Bodley Head, 1965], 166)

43. Nehru, *The Discovery of India*, 522.
44. Ibid. According to Nehru, the Communist Party of India lacked any popular appeal because of its complete lack of understanding and disdain for Indian traditions (ibid.).
45. Ibid., 523.
46. Prakash, *Another Reason*, 225.
47. Ibid.
48. Ibid.
49. The mode in which the nation-state was imagined as gendered in the construction of its modernity in nationalist thought bears noting. For our

purposes, the construction of the nation as feminine and masculine in the respective thought of Savarkar and Nehru in particular is most relevant. Prakash has described "the startling gendering" of the nation as an Indian woman in Nehru's *The Discovery of India* (*Another Reason*, 206). Yet for Nehru, the modernity of the Indian state did not necessary stand for the masculine, inasmuch as that was constructed as the antithesis of the feminine Indian nation. Rather, the masculinity of the Indian state was anticipated in the construction of the Indian nation as feminine; "the 'lady with a past' was a masculine figure in disguise" (ibid., 206–207). In Savarkar's thought, too, the masculine and the feminine coexist in the construction of the nation and the state, although the relationship between the two is configured differently. As in Hindu nationalist discourse at large, the honor of the Indian nation for Savarkar is identified with the honor of Hindu women, both of which were under threat from Muslims. Sarkar notes that "from Savarkar's formative writings on Muslim rule in India, the stereotype of an eternally lustful Muslim male with evil designs on Hindu women has been reiterated and made a part of historical common sense" ("Heroic Women, Mother Goddesses," 185). In this narrative, Hindu women are symbols of the Indian nation, while the Indian nation is conversely imagined as the Hindu mother. The task of Hindu nationalism then becomes that of protecting Hindu women and the Hindu nation. This responsibility is entrusted to Hindu men who are authorized to speak for, and act in the name of, the nation. Savarkar, in *Hindutva*, describes the nation as both the "fatherland" and the "motherland." The duality fits with the roles ascribed to Hindu men by Savarkar. The Indian nation is feminized but the Indian state is masculinized. The foremost obligation of the military-industrial complex and, indeed, of the modern Hindu state itself, is to protect the honor of the Hindu nation.

50. See Dietmar Rothermund, *An Economic History of India: From Pre-Colonial Times to 1991*, 2nd ed. (New York: Routledge, 1993), 124–132, for a description of the thinking of Indian economic planners with regard to industrial and agricultural development in the years leading up to independence and after.

51. To cater to the growing needs of industries over the last fifty years, the Government set up a large number of Industrial Training Institutes, all over the country. It also set up Indian Institutes of Technology, Management Institutes, and Engineering Colleges to train persons with higher management and technical skills...The Government of India set up 48 national laboratories to undertake applied research in chemistry,

physics, electronics, botany etc. (Government of India, Ministry of Labour, "Industrial Development & Progress After Independence," in *Main Report*, vol. 1, chap. 3, *Report of the Second National Commission on Labour*, 2002, http://www.labour.nic.in/lcomm2/2nlc-pdfs/Chap3.pdf [accessed May 11, 2004]).

52. Ashis Nandy, "Introduction," 1.
53. Visvanathan, *A Carnival for Science*, 4. Describing the expectations of science in postcolonial India, Visvanathan points out that

    in the fifties it was like a magic wand...One still remembers the romanticism of Nehru's vision of science. For Nehru, the tryst with destiny was a rendezvous with the modern industrial world and "the future," he proclaimed, "belongs to those who made friends with science." (ibid.)

54. See Nandy, "Introduction," 1–23, and Chatterjee, "Development Planning and the Indian State," 82–103. Scientists and economic planners appointed to advisory bodies set up by the Indian government arguably possessed more political capital than many politicians.
55. The affirmation of discourses of science and technology and the endorsement of the acquisition of scientific and technological skills by the state may not in and of themselves signify anything exceptional about independent India. However, there are two features that distinguish the postcolonial Indian state from other states in this regard. One distinctive feature—in comparison to other postcolonial states—is the attempt to sustain the combination of a secular socialist democracy and an emphasis on technology as twin aspects of a scientific project of nation building. Nanda notes that "for most of the 55 years of independence, especially during the Nehru years, India has tried to combine at least the promise of a secular-liberal cultural modernity along with technological modernization" (*Prophets Facing Backward*, 37). The other distinctive feature marking the Indian experience of science and technology as distinct from that of other states is the fact that as a project of scientific and technological modernity it has inextricably been tied to a project of political modernity, as compellingly described by Prakash in *Another Reason*. "Perhaps nothing is more distinctive about Indian modernity," Prakash comments, "than the intense, highly charged relationship it embodies between science and politics" (11). The historically authorized political salience of science has no doubt contributed to its mobility as metacapital across various fields.

56. David Swartz, "Pierre Bourdieu: The Cultural Transmission of Social Inequality," in *Pierre Bourdieu*, ed. Derek Robbins (London: Sage, 2000), 2: 207.
57. Pamela Shurmer-Smith, *India: Globalization and Change* (London: Arnold, 2000), 37.
58. Ibid., 36.
59. "Only about 2 per cent of the more than 200,000 boys and girls who sit for the JEE every year get through." Shantanu Guha Ray et al., "The Best and the Brightest," *Outlook India*, May 29, 2000, http://www.outlookindia.com/full.asp?fname=cover_story&fodname=20000529&sid=1 (accessed February 11, 2004). The JEE is the Joint Entrance Examination, which is the entrance test for admission to the Indian Institutes of Technology (IITs).
60. Aggarwal, *Documents on Higher Education in India (1781 to 1992)*, 413.
61. Guha Ray et al., "The Best and the Brightest."

    The IITs are the Indian Institutes of Technology, India's most prestigious engineering schools.
62. Shurmer-Smith, *India*, 36.
63. In 2002 India overtook China for the first time as the leading source of international students in the United States. The year 2002 saw 66,836 students from India enrolled in academic institutions in the United States as compared to 63,211 from China. Institute of International Education, "International Student Enrollment in U.S. Rose 6.4% in 2001/2002, Matching Last Year's Increase as highest growth since 1980—India is the Top Sending Country," *IIE Online*, November 18, 2002, http://www.iie.org/PrinterTemplate.cfm?Section=News_Announcements&template=/ContentManagement/ContentDisplay.cfm&ContentID=5470 (accessed June 24, 2004).
64. Shurmer-Smith, *India*, 35–36.
65. Ibid., 36.
66. Ibid., 36, emphasis in original.
67. Madhulika S. Khandelwal, "Indian Immigrants in Queens, New York City: Patterns of Spatial Concentration and Distribution, 1965–1990," in *Nation and Migration: The Politics of Space in the South Asian Diaspora*, ed. Peter Van der Veer (Philadelphia: University of Pennsylvania Press, 1995), 179. The reference is to the Immigration and Naturalization Service in the United States.
68. Bourdieu, *Pascalian Meditations*, 96–97.
69. In chapter 3, I defined *scientificity* as the abstract signifier of the values of reason, enlightenment, and rationality and *technicity* as the pragmatic

application of these values in concrete contexts. The nomos of the technological field in India is marked by a divide between the two, but the equation between them is recalibrated in different phases. The term "technicity" should not be confused with Nandy's use of "technicism" identified earlier in this chapter.

70. While some of the work referred to in this section uses the term *class* in a Marxist sense, as will be evident, it is nonetheless compatible with Bourdieu's framework.
71. Vanaik, "Consumerism and New Classes in India," 227–234.
72. Ibid., 228. Vanaik points out that these elites are often "misdescribed" as the Indian middle classes and that the so-called Indian middle classes are not analogous to middle classes in Western nations, which are genuine median groups. He states that the

    > Indian middle class really means the top 20–25 per cent of the population with three fourths or more below it...it is really a part of a dominant elite made up of professionals and various kinds of property/capital owners. Even if this elite was not 20–25 per cent but 10–15 per cent or only 5–10 per cent of the total population, it would, in absolute numbers still be very sizeable. (ibid., 228–229)

    Stating that the Indian "middle class" is a "fashionable category" gleaned from consumption and lifestyle indicators, Harris-White notes that this group "swelled from an estimated 60–80 million urban people in the mid-1980s to a (probably overestimated) figure of about 300 million in the mid-1990s." Barbara Harris-White, *India Working: Essays on Society and Economy* (Cambridge, U.K.: Cambridge University Press, 2003), 61.
73. Harris-White, *India Working*, 1.
74. Ibid.
75. As of the 2001 census, India's population was 1028.7 million. Government of India. Office of the Registrar General and Census Commissioner, India, "Census Data 2001—National Summary Data Page (NSDP)," *Census of India*, n.d., http://www.censusindia.gov.in/Census_Data_2001/National_Summary/National_Summary_DataPage.aspx (accessed June 22, 2008). The total urban population was approximately 286.12 million. See idem, "Urban Population," *Census of India*, n.d., http://www.censusindia.gov.in/Census_Data_2001/Census_data_finder/A_Series/Urban_population.htm (accessed June 22, 2008).

    As per the 2001 census, a little over a fifth of India's urban population lives in slums. In the 2001 census, the figure for "Total Slum Population returned as per provisional results of Census of India 2001 was 40,605,418, comprising of 22.76% of the total urban population of the cities/towns reporting slums." This was based on criteria for defining a slum population. See idem, "Census News—Data Highlights—New Releases," *eCENSUSIndia*, no. 4, 2001, November 30, 2001 [bulletin], http://www.censusindia.gov.in/Census_Data_2001/Census_Newsletters/Newsletter_Links/eci_4.htm (accessed June 22, 2008). The entire nonslum population does not necessarily qualify as elite, but estimates about the size of Indian elites vary widely.

The fraction of India's population that can speak English also varies according to different estimates. Crystal notes the results of a survey that suggest that about 350 million Indian speakers "had the ability to carry on a conversation in English." The author describes this finding "as an amazing increase over the estimates of the 1980s, when only about 4%–5% of the population were thought to use the language." David Crystal, "Subcontinent Raises Its Voice," *Guardian Weekly*, November 19, 2004 http://education.guardian.co.uk/tefl/story/0,5500,1355064,00.html (accessed June 12, 2005). The figure of 4–5 percent may be more relevant for estimating the proportion of the population that is fluent in English. Kumar notes that in urban India 18 percent of the population is fluent in English, but in India as a whole the number of English speakers is very low—less than 1 percent as per a 1999 national survey. Sanjay Kumar, "India's Urban Middle Class," in *Values and Life Styles in Urban Asia: A Cross-Cultural Analysis and Sourcebook Based on the AsiaBarometer Survey of 2003*, ed. Takashi Inoguchi, Miguel Basanez, Akihiko Tanaka, and Timur Dadabaev (Mexico City: SigloXXI Editores, 2005), 168.

76. Pranab K. Bardhan, *The Political Economy of Development in India* (New York: Basil Blackwell, 1984).
77. Ibid., 32.
78. Khilnani, *The Idea of India*, 147.
79. Ibid., emphasis added.
80. Bose, "'Hindu Nationalism' and the Crisis of the Indian State," 132.
81. Ibid.
82. Ibid.
83. Ibid.
84. Ibid.

Chapter 5

The Technocultural Hindu Nationalism of Globalized India

This chapter follows the development of the Indian technological field and technological capital since 1991, with reference to two seminal events that have profoundly influenced the character of Indian social existence in the last two decades: (a) the liberalization and globalization of the Indian economy in 1991 and (b) the rise of Hindu nationalism as a strong political and cultural movement on the national stage in the 1990s. Each of these paradigms proposes a particular definition of the Indian nation and a blueprint for national progress that will help realize India's potential in the globalized world of the twenty-first century. The neoliberal paradigm of governance adopted in 1991 articulates itself in fundamentally economic terms, while the agenda of Hindu nationalism is expressed primarily in cultural terms. The domain of technology, especially the Internet, is a key area where these respective perspectives on the Indian nation converge. This convergence produces, as much as it reflects, a new relationship between technology and nationalism—which I term "technocultural Hindu nationalism"—that typically finds expression on the Internet in the form of Web sites addressed at a global Hindu audience. An examination of the portal *Hindu Universe* illustrates some of the key characteristics of technocultural Hindu nationalism.

A comprehensive and multipurpose portal aimed at a global Hindu population, *Hindu Universe* is a particularly good example for reading technocultural Hindu nationalism. The portal consists of an umbrella structure incorporating numerous categories of content, community, and commerce. It can be conceptualized as a part of a complex transnational network, virtual and offline, of Hindu organizations. It includes, and links to, a wide range of offerings—from channels on health and lifestyle, marriage and family, and education to a record of community events; from an array of Hindu-themed shopping options to user-generated content and interactive features; from news and information to digital greeting cards for Hindu festivals. Table 1 provides a listing of some of the sections of the *Hindu Universe* portal, as well as some of the related sites linked to and through it, as reflected on the home page of the Web site.[1]

Hindu Universe is analyzed here as a dense network of discursive objects or "texts" located in economies of technology and culture that are at once economies of power. It can be conceptualized as standing at the confluence of multiple discourses that it simultaneously embodies: of national and diasporic identity, capital and community, nationalistic fervor and religious sentiment, instrumental practices and ideological commitment, and material consumption and symbolic and performative affirmations. The portal may also be seen as a space marked by the logics of different Indian fields—such as the cultural, technological, and economic—and the exchange rates between them in the era of global capitalism and resurgent Hindu nationalism. This methodological approach is helpful in allowing an identification of the negotiations between technology and majoritarian nationalism as they play out in cyberspace as well as pointing to what they share with "offline" Hindu nationalism. Any method, however, constrains as much as it enables; the limitations of the approach followed here should be noted. The perspective adopted here calls for significant attention to the overlap and exchange between the technological, cultural, and economic fields in India, which necessarily involves delving into the history of these fields. This focus precludes a detailed close reading of each Web site section or of several of the sites linked through *Hindu Universe*.[2] This methodological limitation is sought to be offset by the statistical and exploratory contextual analysis offered in chapter 6, which presents the spread of themes in Hindu nationalist discourse (as well as Sikh nationalist and Kashmiri nationalist Web sites).

TABLE 1. *Hindu Universe*: sections and linked sites.

Section/Category Name	URL
Arts	http://www.hindunet.org/arts/
Gods, Sages, and Gurus	http://www.hindunet.org/god_gurus/
Scriptures	http://www.hindunet.org/scriptures/
Internet Books and Resources	http://www.hindunet.org/hindu_resources/
Temples and Organizations	http://www.hindunet.org/hindu_organizations/
Interfaith Relations	http://www.hindunet.org/interfaith_relations/
Dharma & Philosophy	http://www.hindunet.org/dharma/
Worship	http://www.hindunet.org/worship/
Customs	http://www.hindunet.org/hindu_customs/
Sciences	http://www.hindunet.org/sciences/
Social and Contemporary Issues	http://www.hindunet.org/issues
History	http://www.hindunet.org/hindu_history/
Channels	
Resources	http://www.hindunet.org/resources/
Kids and Youth	http://www.hindunet.org/kidsyouth/
Women	http://www.hindunet.org/women/
Marriage and Family	http://www.hindunet.org/marriagefamily/
Pujas and Samskars	http://www.hindunet.org/pujassamskars/

(continued on next page)

TABLE 1. *(continued)*

Section/Category Name	URL
Spiritual Tourism	http://www.hindunet.org/spiritualtourism/
Community	http://www.hindunet.org/communityactivism/
India	http://www.hindunet.org/india/
User Created	
Sites	http://www.hindunet.org/memberscgi/webdirectory.cgi
Photoalbum	http://www.hindunet.com/memberscgi/communityalbum.cgi
Clubs	http://www.hindunet.com/memberscgi/clubdirectory.cgi
Related Sites Linked from the Home Page	
Visit 3000 Temples Online	http://www.mandirnet.org/
Listen to Bhajans, Kirtans…	http://www.archaka.com/
Puja Archaka…	http://www.archaka.com/
Biographies of Great Hindus	http://www.freeindia.org/biographies/
HinduGifts	http://www.hindugifts.com/
HinduNet Signature Merchandise	http://www.cafepress.com/hindunet2/
Jyotishweb—Astrology	http://www.jyotishweb.org/muhurath/home.asp
Gita Today	http://www.gitatoday.com/
HinduLinks Universe	http://www.hindulinks.org/
Upcoming Events	
This Week	http://demo.ghen.net/ems/events.php?mode=thisweek
This Month	http://demo.ghen.net/ems/events.php?mode=thismonth
This Year	http://demo.ghen.net/ems/events.php?mode=thisyear

On *Hindu Universe* and related sites accessible through it, two claims about Indian and Hindu identity are to be found, juxtaposed against and jostling with each other. In the first proposition, Hinduism is presented as a religion that is fully compatible with a technologically and economically interconnected globe. The implication is that Hindus can participate and flourish in the world of today precisely because they are culturally grounded in a universalist Hindu ethos and heritage that are uniquely attuned to a global marketplace. The value of Hinduism as a source of economic tools, skills, or solutions is explained in terms of a more foundational cultural universalism found in the Hindu faith. The second proposition, consistent with Hindu nationalist ideology, is that "Indic" identity—the sum of religions, sects, and practices rooted in the indigenous Hindu dharma of the subcontinent—equals and exhausts Indian identity, the former granting the latter its essential character. Consequently, non-Indic identities, such as those of the Indian Muslim or Christian, do not count as Indian in the cultural sense, even though they may qualify as Indian by virtue of citizenship.

Hindu Universe offers online visitors access to a range of social activities, cultural resources, and commercial products. A visitor to the site can avail of services such as the chat facility; participate in one of several discussion forums on topics ranging from secularism to Kashmir and from the Indian education system to the Indian Air Force; send e-cards for festivals such as Pongal or Baisakhi; read up on the Hindu sacred texts, philosophy, and mythology; or shop for jewelry, astrological reports, or school bags. The portal is part of a cross-linked, and sometimes confusing, virtual global network of Hindu associations and organizations. It is described as "the website for GHEN (Global Hindu Electronic Networks). GHEN is one of the many projects undertaken by HSC (Hindu Students Council)."[3] The Hindu Students Council, in turn, is defined as a "voluntary run organization committed to realizing the ancient Vedanta truths such as *Vasudaiva Katumbakum (The Whole World is One Family)*."[4] On the Web site of the Hindu Students Council, the organization is described as "an international forum that provides opportunities to learn about Hindu heritage through various activities, events and projects. HSC presents ample opportunities for self-development at the spiritual and professional level."[5] The site also proclaims that "HSC represents the first and only North American and international attempt by students and young professionals like yourself to explore, discover & experience the immense treasures of the time-tested knowledge and wisdom of the great Hindu culture."[6] The text strongly resembles advertisements for new age techniques for acquiring management skills, where the truths of Vedanta are valued as much for their instrumental contribution to career enhancement as for their spiritual value. If the world is a family, it is the family of late capitalism where the *sine qua non* for membership is the possession of the professional skills required to survive in the global marketplace.

Upon browsing through the Hindu Students Council Web site, it becomes clear that all the organization chapters are located in North America.[7] At the very least, there are two assertions on the site that are problematic both as empirical claims and as political arguments: first, the definition of North America as synonymous with "international," given that various populations of the 20-million-strong Indian diaspora are to be found in virtually every country on the world map, and second, the fact that an organization based exclusively in North America, where Hindus number less than 2 million, should arrogate for itself the right to speak on behalf of hundreds of millions of Hindus worldwide.[8] In ignorance of the historical and sociological realities of the lives of Hindus across the globe, the Hindu Students Council Web site proposes a normative and prescriptive model of Hinduism as a Brahmanical religion rooted in the authority of the Vedas.[9] The claims of global reach and universal significance, then, are shallow but strategic, reflecting an identification with North America as the center of the globe and with upper-caste Hinduism as the most authentic incarnation of the Hindu faith.

On the same portal, the terms *Hindu* and *Hinduism* acquire new meanings in the realm of commerce as well. *Hindu Universe* includes a link to a commercial Web site, *Hindugifts*, that sells everything from gift vouchers to sarees and from digital cameras to phones.[10] The site offers shoppers the choice to gift to India or globally, and its products are priced in both Indian rupees and American dollars. Another site linked from the *Hindu Universe* portal is the *Hindunet Signature Merchandise* Web site, which offers its viewers "Hindu-themed" baseball jerseys and mouse pads with messages such as "Proud to be a Hindu" and "The entire universe is one family."[11] The site appears targeted primarily at a diasporic, especially U.S.-based, customer base of Hindus since (a) the prices of products are in dollars and (b) some of the products, such as baseball caps, would have relatively low market demand in India, given that baseball is neither played nor widely followed in the country.

Both sites reflect the marketing of Hinduism as a globally saleable product. They define the act of economic purchase as a cultural obligation to Hinduism. The activity of consumption, actively encouraged, is itself branded as Hindu. Participation in a chat room discussion about Hindu culture or history on *Hindu Universe*, followed by the act of purchasing a "Hindu" calling card through the same portal, allows an economic transaction to be recast as a more fundamental act of cultural loyalty or duty. Likewise, the *Mobilehindu* Web site, which is also part of the *Hindu Universe* network, sells U.S.-based viewers cellular phone subscription packages from telecommunications service providers like Nokia, Cingular, or Verizon, much like any other e-commerce gateway.[12] Yet the entire economic transaction is marked as an exchange between Hindus. The Hinduism of the *Hindu Universe* portal is completely in tune with the requirements of global capitalism.

At first glance, the home page of *Hindu Universe* does not appear to be offensive in its views on Islam and Christianity. Yet its acts of commission and omission make its ideological affinities clear. For example, the section titled "Dharma & Philosophy" on the home page of the portal offers an "Intro to Hindu, Jain, Buddhist, Sikh Dharma and Philosophy" with no mention of Islam or Christianity or, for that matter, Zoroastrianism or Judaism.[13] Neither does there appear to be any material on Indian syncretic religious practices. One has to but dig slightly deeper and the anti-Muslim and anti-Christian bias of the portal becomes visible—for example, in the section on "Interfaith Relations."[14] Here, a category of content titled "Conversions" speaks of Hindus converted to Christianity and Islam by force and deception.[15] The portal section on Hindu-Muslim relations posits that the "Moslem invasion of India has been one of the most brutal events in the history of mankind. During the Moslem [*sic*] rule, millions of Hindus were massacred and converted by force."[16] The portal also includes a "History" section, which contains numerous articles espousing anti-Muslim and anti-Christian sentiments.[17] In keeping with the claim that Hindu dharma provides the overarching structure of Indian identity, the articles define Islam and Christianity as alien religions foisted on to India by Islamic invaders and European colonizers.[18]

The *Hindu Universe* portal signifies the confluence of a new role for technology in the current global capitalist economy with assertions of cultural nationalist sentiment. Each dimension—the technological and the cultural nationalist—influences the mode in which the other is articulated. Hindu culture is emphasized as scientific, for example, in the claim about the universal scientific truth of Vedanta.[19] Correspondingly, acts of technological usage, such as purchasing an item through an e-commerce Web site, are branded as Hindu activities. Not all visitors to such sites may sympathize with Hindutva ideology; some may only visit the Web site to shop for a festive occasion. And content on these sites does not exhaust the views of a diasporic Hindu community. But the features of the online discourse described previously—for example, the assertion that North American Hinduism is the seat of international Hinduism, the happy marriage between the Hinduism of the Global Hindu Electronic Networks Web sites and global capitalism, and the antiminority slant of such online Hindu discourse despite its professed inclusivity and universalism—are not the products or artifacts of happenstance. Technocultural Hindu nationalism may be viewed as the expression of a new form of metacapital, whose circulation is made largely possible by, and whose value is largely dependent on, the changes within various fields (technological, cultural, economic, and educational) and the equations for exchange of capital between these fields since 1991. Relatedly, the Indian technological class has also experienced internal change since 1991 with regard to its composition in terms of caste, income level, and type of educational qualifications. The technocultural Hindu nationalists, the possessors of this new type of technological-cultural capital, fit a particular profile within the broader technological class. They also map on to a particular profile within the broader category of those who support Hindu nationalism.

In describing technocultural Hindu nationalism, this chapter proceeds as follows. I first describe the liberalization of the Indian economy as an embrace of neoliberalism by the Indian state. The impact of this change is a foundational redefinition of scientificity, as a form of scientific-technological capital, at the level of the nomos of the state toward the objective of optimizing this species of capital for circulation in a national-global capitalist economy. In the next section, I describe the changes in the technological field since 1991, with a focus on the increasing importance of the Internet in Indian society. I show how the Indian state's emphasis on information technology in a globalized world, symbolized by the Internet, further alters notions of scientificity along technocratic lines and impacts various fields and species of capital. The third section traces the rise of Hindu nationalism as another movement—much like economic liberalization—that seeks to radically recast the nomos of the Indian state and, through it, all fields in Indian social space. The chapter concludes with a description of technocultural Hindu nationalism as a selective reworking of the neoliberal technocratic nomos of the Indian state within the cultural nationalist framework of Hindutva.

The Neoliberal Turn of the Indian State

Theorizing Neoliberalism

Bourdieu presents an analysis of neoliberalism as doxa—an unquestionable orthodoxy that operates as if it were the objective truth across social space, from the practices and perceptions of individuals (at the level of habitus) to the practices and perceptions of the state and social groups (at the level of fields).[20] Neoliberalism insists on a mathematical model of economic behavior as the paradigm for human sociality, discarding the irreducibly social that does not translate into mathematical terms.[21] Presenting itself as the "scientific description of reality," neoliberal discourse does not recognize historically or socially constituted rationality.[22]

The status of neoliberalism as doxa, Bourdieu tells us, is "what gives the dominant discourse its strength."[23] The notion of neoliberalism as the self-evident truth about the social has been steadily prepared over decades in France and the United Kingdom by partisan groups of academics, media persons, and businessmen, and has been routinely reinforced by ordinary citizens and the media.[24] As a result, in France, the state has begun to abdicate its role as the protector of social benefits in education, health, and welfare.[25] In the United Kingdom and the United States, permanent jobs with benefits are being replaced by temporary and underpaid jobs.[26] Neoliberalism also erodes the autonomy of the arts by subjecting artistic production to market pressures.[27] Perhaps the most devastating impact of neoliberalism has been that it has rendered ineffective the collective institutions that could offer any resistance to it, foremost among which is the state itself.[28]

What the neoliberal worldview actually achieves is nothing other than the oldest dream of capitalism—the establishment of a framework for the accumulation and distribution of profit according to Darwinian principles. Bourdieu uses the term *state tradition* to describe the role of the state as guarantor of the welfare of its citizens, and points out that where state tradition has historically been the strongest, resistance to neoliberal doctrine is correspondingly stronger.[29] Where state tradition has been historically weak, such as in the United States, the state is most visible in its policing function.[30] While the state usually does not possess complete autonomy, the older the state "and the greater the social advances it has incorporated, the more autonomous it is."[31] The neoliberal state closely resembles the model of weak state tradition, with a minimal role in providing social benefits and greater license to concentrate on its policing function.

Bourdieu argues that wherever neoliberalism has become the nomos of the state, an effort has been made to erase the history of the state. Such efforts seek to regress the state to the earliest stages of its development when its role was restricted largely to its policing function and its economic power.[32] The nomos of neoliberalism, then, relies fundamentally upon the assumption that social existence is not historically constituted and that it is not worthwhile to invoke historical precedent in understanding the present. The doxa of neoliberalism is enforced by the transnational networks of globalization, albeit through the agency of neoliberal states. Neoliberalism operates as the doxa of a national habitus, restructuring fields and relations between fields. At both the micro level and the macro level, it establishes itself as a set of truths and a way of looking at the world.

The Liberalization of the Indian Economy and the Shift in the Nomos of the Indian State

The Indian state took the decision to liberalize the economy in 1991 in response to a balance-of-payments crisis catalyzed by the first Gulf War. The situation was characterized by depleting foreign exchange reserves, falling exports, soaring inflation, and drying up of workers' remittances. The reforms included both short-term and long-term measures; then finance minister Manmohan Singh "went beyond the traditional stabilization package, announcing a program of economic reforms that constituted a significant reversal of some of the most egregious aspects of earlier policies of regulation and government intervention in the economy."[33] The measures included reducing the fiscal deficit; devaluing the rupee; opening sectors of the economy to private investment, both domestic and foreign; and removing government controls in foreign trade.

The impact of the reforms is a topic of contentious debate, and undertaking a detailed assessment of the same is beyond the scope of the discussion here. For our purposes, the point to be noted is that the conversation in much mainstream English-language public discourse in India since 1991 reflects the viewpoint that the irreversible tide of liberalization cannot and should not be resisted, despite due acknowledgment of the severe inequalities that afflict Indian society. There also appears to exist a consistency among professional, technical, and specialized arguments among large sections of policymakers and academics about the inevitability of liberalization and globalization.[34] As the Indian economy continues to grow at around 8 to 9 percent a year, the consensus appears to get further entrenched.

The actions and rhetoric of numerous Indian state and nonstate agencies also seem to endorse globalization and liberalization as desirable transformative forces that will ultimately provide not only economic rewards, such as increased global competitiveness of Indian companies and healthier foreign exchange reserves, but also clear social benefits, such as more job opportunities and egalitarian distribution of wealth. Most Indian political parties, including the BJP, Congress, and even some Left parties, are generally in favor of the reforms despite professed differences on specific issues. Liberalization and globalization are now invested with the burden of fulfilling the promise of postcolonial modernity that socialism has sorely failed to deliver.[35]

My point here is not that India before 1991 was necessarily a viable socialist economy. Nor am I suggesting that the impact of the reforms has been unmistakably negative. My argument, rather, is that the discourse of justification, according to which policy changes and political decisions are explained to an Indian public, has undergone a reconfiguration, where the old arguments about the moral superiority of socialism over capitalism and the quasi-Marxist grounding of economic policy in a narrative of social justice are no longer required. After independence, the value of scientificity as a highly desired form of capital derived from its value within and across fields (economic, educational, cultural, etc.) and its endorsement by the Indian state. With the neoliberal turn in 1991, scientificity itself is not rejected as a principle of generating capital in the vision of the Indian state (to the contrary, scientificity even gains in value as a form of capital in neoliberal India). The rhetoric of national progress is also preserved in the endorsement of technology by the Indian state. But the structure through which progress will be achieved is understood in radically different terms from the Nehruvian socialist model.

This is the critical change in the nomos of the state (or the doxa of the national habitus) brought about by the adoption of a neoliberal framework by the Indian state. The value of scientificity is delinked from any direct socialist imperative—economic or political. It is not expected that scientificity should lead to national self-sufficiency in the Nehruvian-era definition of the term—that is, as an independence from necessary economic and political dependency on other nation-states. Instead, scientificity is redefined in the context of revenue-earning capacity in a global economy. At the same time, the socialist framework is no longer considered the appropriate context in which scientificity can operate optimally. The quality of scientificity is now best thought to flourish in a free market environment with a much greater role for the private sector. And after 1991, scientificity as a principle of state vision is not burdened anymore with having to inculcate in its possessors the qualities of a critical, enlightened, and rational—in a word, "scientific"—temper and outlook. As we will shortly see, in Hindu nationalist discourse, scientificity is also reclaimed as an essentially Hindu quality that is ideal for a globalized, economically and technologically interconnected world.

The Technological Field and Class in India Since 1991

Information Technology and the Internet in India After 1991

In a post-liberalization world, India's value for the global economy depends largely on its IT capabilities and offerings. Sachs, Varshney, and Bajpai point out that "India is becoming one of the most important players of the world in the IT sector and it is the fastest growing foreign exchange earner for the country."[36] Greenspan observes that "inside India, there is no doubt that the IT industry has created jobs, strengthened exports and made substantial contributions to economic growth."[37] Internet-related initiatives in computing software and hardware, telecommunications, and business constitute an important segment of the Indian IT industry. Wolcott and Goodman argue that "the Internet is central to the new vision of India as an IT power in many respects."[38] Sachs, Varshney, and Bajpai list the Internet as one of the key factors that attract American and European companies to set up back-office operations in Indian metros.[39] In fact, the Internet seems to share a symbiotic relationship with the economic liberalization of India. As Wolcott and Goodman note, "In the latter half of the 1990s, the pace of reform quickened considerably, with the Internet playing an integral role, both as a factor driving reform and as a beneficiary of change."[40] Domestically, the Internet is the key driver of PC sales.[41]

The central government, numerous state governments, industry, policymakers, media, and civil society organizations seem to agree that information technology and the Internet have tremendous potential for the economic and social development of India. Several Indian states, from Andhra Pradesh and Kerala in the south of India to Tripura in the northeast, have implemented ambitious e-literacy and e-governance initiatives. National and state government projects are in the process of bringing the benefits of information technology and the Internet to rural Indian communities, with the objective of promoting socioeconomic development. For example, the Warana village experiment, jointly undertaken by the national government, the Maharashtra state government, and the Warana Vibhag Shikshan Mandal (WVSM), aims to effectively use information and communication technologies, including the Internet, for the socioeconomic development of seventy villages near Warana Nagar in Maharashtra.[42]

The Indian state recognizes that Indian technology professionals constitute an immensely valuable resource of human capital for the Indian nation. Greenspan points out that "Silicon Valley is said to have a 300,000 strong Indian techie community, with Indian-owned or backed firms valued at US$ 40 billion."[43] She notes that "by 2003, the phenomenal success of these multimillionaires, alongside the thousands of programmers and engineers who were working at all levels of the industry, ensured that Indian IT had established itself at the core of global IT economy."[44] The policies and actions of the Indian state since the 1980s have sought to create a context that can promote domestic talent in the arena of technology while also cultivating the U.S.-based community of Indian technology professionals, with a view to incorporating their contributions toward national development. As Chakravartty notes,

> In the past twenty years, Indian national and regional governments have embraced, enticed, and tried to formalize networks with non-resident Indians in high-tech sectors, particularly in the United States (McDowell 1997). These sectors include telecommunications, information technology, and electronic commerce...Since the mid-1980s, the Indian state has rapidly expanded its technological

> infrastructure and promoted the nation's vast pool of highly trained low-wage labor, alongside the unmet market potential of its growing middle class in order to attract foreign investment.[45]

With regard to domestic IT-related initiatives, Singhal and Rogers speak of the emergence of nascent Indian "technopolises," defining the term as "a geographically concentrated high-technology complex characterized by a large number of entrepreneurial spin-off companies."[46] The technopolis, the model par excellence of which is Silicon Valley, is at once a center for technological innovation and for wealth creation. National and state governments promote the cities of Bangalore and Hyderabad as Indian versions of the famed American technology hub.

As a result of these developments, the Indian technological, economic, and educational fields have been significantly restructured since 1991, as have the equations of exchange between them. The Indian state now encourages its subjects to acquire global capital, especially in technology-related fields, whether in India or overseas. The development of the World Wide Web in the early 1990s, the Internet boom in Silicon Valley in the late 1990s, and the ensuing demand in the United States for Indian programmers have meant that the conversion rate of technological to economic capital rises dramatically. The educational field in India has been visibly impacted by changes since 1991, with the curricular emphasis slowly but unmistakably having shifted toward the acquisition of "skills" needed to be competitive in a global economy. Indian technological institutes, including engineering colleges, polytechnics, and computer-training centers, by one account, produce a million professionals every year.[47] Numerous Indian universities, especially in the metros, are increasingly offering courses in information technology to feed growing demand.

These developments, combined with the global demand for Indian technology professionals, have meant that the exchange rates between educational capital, economic capital, and cultural capital have also changed since liberalization. Training at state institutes in technology skills or at private institutes offering short-term courses, for example, promise the kinds of job opportunities in India and overseas hitherto available only to engineering graduates. In addition to substantial economic capital, such qualifications also promise the prestige of an international career and the accompanying cultural capital. These changes also point to the struggles over educational, economic, and cultural capital in India. A wider cross section of Indian society is now staking its claim to both increasing their capital in these respective fields and to altering the nomos of the educational and economic fields. As Friedman notes,

> Hundreds of thousands of young Indians mostly from lower-middle-class families suddenly have social mobility, motor scooters and apartments after going to technical colleges and joining the Indian software and engineering firms providing back-room support and research for the world's biggest firms—thanks to globalization.[48]

Keniston has described the emergence of a new elite group in India comprised of workers in high-tech professions, "the beneficiaries of the enormously successful information technology (IT) industry and the other knowledge-based sectors of the economy, such as biotechnology and pharmacology."[49] Keniston identifies the characteristics of this "digerati" as affluence, cosmopolitanism, and global mobility.[50] In one respect, the digerati are truly an elite within an elite. Their salaries "are relatively low by Western standards, but with annual salary growth rates of over 20 per cent for the last 5 or 10 years, far above those of their otherwise equally educated classmates in India," such as those in the natural sciences, mechanical engineering, or chemical engineering.[51] In another respect, they are distinctly different from old Indian elites:

> the privileges of the new digerati are based not on caste, inherited wealth, family connections or access to traditional rulers, but on a combination of education, brainpower, special entrepreneurial skills, and ability to stay on the "cutting edge" of knowledge.[52]

Khilnani describes a similarly globalized class of Indian professionals that has emerged in India since 1991 and is typified today by the young software programmer or MBA. He argues that this class is "not constrained by the territorial frame of the nation state."[53] For the members of this class, "India is merely one stopping place in a global employment market."[54] The movement and composition of the class is determined by two closely related factors: direct global demand; and upward economic and social mobility within the internal class structures of Indian society, made possible by increased educational and economic opportunities. However, despite the unarguably important role played by global forces in shaping the composition and dynamics of the new, post-1991, technological class, it is not the case that the Indian nation has become redundant in the equation. Indeed, new conceptions of national affiliation have emerged in liberalized India. The globetrotting software programmer is the new model of responsible citizenship who, by choice of vocation, expresses his or her patriotic commitments.

The Rise of the Indian Diaspora

The change in notions of national affiliation is strongly related to the increasing importance of NRIs, especially those based in the United States, for the Indian state since liberalization. Chakravartty argues that NRIs embody a form of "flexible citizenship" wherein "emigration from India to the United States is not understood as a permanent departure to another nation-state."[55] The new importance of NRIs derives from their crucial role in

> legitimizing a relatively new technocratic development agenda in India. This development agenda, initiated in the 1980s and codified in the 1990s, appeals to the symbolic power of postcolonial technological nationalism with a new emphasis on the meritocratic logic of the marketplace.[56]

The key characteristics of the model of flexible citizenship, at once transnationalist and capitalist, are the following: an emphasis on an entrepreneurial framework for generating scientific knowledge in contrast to the Nehruvian model of statist commitment to science; a nationalism that is "pro-nation but anti-state" and, at the same time, corporatist and inspired by American business culture; and an ambivalent relationship toward Indian society where, likewise, state action for fulfilling social commitments is rejected in lieu of private contributions toward any such responsibilities.[57] Muppidi argues that in liberalized India the NRI has been inscribed in state discourse as the most authentic incarnation of postcolonial citizenship, since the NRI appears optimally located between the desired modernity of the West and an essential Indianness.[58]

Greenspan also notes that "for decades India's diasporic population has served to inspire their local counterparts," citing as an example an article in a business magazine that glorifies the achievements of diasporic Indian capitalism.[59] While the nomos of the technological and other fields has been profoundly altered since 1991, the general notion that an occupation in technology is an act of service to the nation continues to endure in the Indian social imaginary, both within India and in the diaspora. The idea continues to flourish, even as the concept of serving the nation takes on new ideological meanings.

Diasporic and National Elites in the Technological Field

The rhetoric of the benefits of technology, however, immediately begs the question of what proportion of the Indian population actually stands to benefit from information technology and the Internet. Access to information technology and the Internet in India is very much the privilege of an elite. Greenspan notes that "[n]owhere is the digital divide more intensely felt than in India, a country that—at least when viewed from the prism of IT—is entering the 21st century radically fractured from within."[60] For Indians who do not speak English, "the barriers to the Information Age are almost insuperable," given that the majority of Web sites are in English and there is hardly any software in non-English languages.[61] Likewise, the impact of the Internet in terms of usage is limited compared to its instrumental value in driving economic reforms. By February 2005, the total number of phone subscribers had increased to 97.03 million.[62] Yet, as of December 2005, India had only 5.56 million Internet subscribers.[63] The number of Internet users in India as of 2006 is just 40 million, a fraction of its population of more than one billion.[64]

Wolcott and Goodman note that, "while in the context of the entire country, usage of the Internet remains low," the Internet usage base has grown at a fast pace since 1995, when commercial Internet service was first introduced in India.[65] The demand for Internet usage is very high, especially among the urban, educated, middle-class population.[66] All major Indian cities—more than 450—are networked, but rural areas suffer from negligible access.[67] Internet usage is limited to an "urban core," and the predominant users of the Indian Internet are the educated middle classes.[68] A large number of Indian expatriates, such as professionals and students in the United States, are a key segment for promoting the growth of Internet usage.[69] Almost 90 percent of all Web sites visited by Indians are in the English language.[70] As high a figure as 60 percent of Indians use cybercafés for Internet access, which numbered over twelve thousand as of June 2002.[71] The high rate of cybercafé use is explained by superior connectivity and falling rates for hourly usage, as well as by the relatively high cost of using the Internet from home.[72] According to one calculation, only 1.6 percent of Indians can afford to spend US$30 a month on telecommunications.[73]

It is helpful here to profile the theoretical class of *Internet beneficiaries* to get a sense of the social segments that stand to gain the most from the technology-oriented policies of the Indian state, given the centrality of the Internet to IT development since 1991. The category of *Internet beneficiaries* is used here to distinguish members of this class from the wider category of *all Internet users*, which may include very infrequent users as well. The class of Internet beneficiaries can be conceptualized as a subset of the broader technological class. The Indian technological class was defined as the theoretical class structurally most privileged with regard to the Indian technological field. At the core of the technological class is an urban, specifically metropolitan, English-speaking population, with the highest amount of economic and cultural capital. This is followed by a secondary tier that is urban and metropolitan and that possesses economic capital but not quite the same level of cultural capital as the core population. Finally, the class includes a third tier, possessing a modicum of economic capital through location in an urban, but nonmetropolitan, economy. Drawn from the core first tier, the social groups most likely to qualify as the prime beneficiaries in the technological field—and so most likely to set the stakes or define the nomos of the field—are urban Indian upper-caste and upper-class professional segments, as contrasted to both urban industrial elites and rural agrarian elites.

Based on the facts noted earlier about the English-language bias of the Internet and the urban skew of Internet access, the class of Indian Internet beneficiaries would span Indians who are "prosperous, urban, highly educated and concentrated in technical fields" as well as U.S.-based Indians.[74] This class of Internet beneficiaries can be envisioned as consisting of different segments or fractions of a broad elite that occupy *allied positions* in the technological field, but *opposing positions* in other fields. One component is a diasporic, U.S.-based, population of high-tech entrepreneurs and workers in India who possess high amounts of educational, economic, and cultural capital. A second component is the Indian-based "digerati" mentioned earlier, who also possess a healthy combination of educational, economic, and cultural capital, though perhaps somewhat less than their U.S.-based counterparts. Following Keniston's description about the relative privilege of the digerati over their equally well-educated nondigerati colleagues, a third component of the IT/Internet class would be those Indians qualified and working in science and engineering (but not in high-tech industries) who possess a somewhat lower amount of economic and cultural capital. Each of these three segments can further be divided into at least two categories: one, those who have also been traditionally privileged in terms of caste, education, and socioeconomic status—that is, in terms of cultural, educational, and economic capital; and two, those who in terms of internal class and caste dynamics have historically occupied a *relatively* subordinate position among the Indian elite and are now challenging the dominance of an upper-caste, upper-class professional elite in the technological field and other fields. The picture that emerges of the IT/Internet class is that of a fractured combination of old and new segments of a broad Indian elite, located across the globe (largely in India and the United States). Each segment possesses a minimum amount of a combination of educational, economic, and cultural capital—such as an undergraduate degree, basic fluency in English, and the financial capacity to absorb the cost of an Internet connection or cybercafé charges—although there would be significant variation in the combination and amounts of capital across the members of each fraction. What all segments share, however, is a de facto ideological investment in the neoliberal paradigm and an agreement about the need for the Indian state to focus on high-tech development. Another movement, Hindu nationalism, to which I now turn, has similarly sought, especially in the 1990s, to refashion the nomos of the Indian state.

The Resurgence of Hindu Nationalism

Contemporary Hindu nationalism is associated with a group of related organizations often referred to as the Sangh Parivar. The BJP is the political face of the Sangh group, while other groups define themselves as cultural or social organizations. The Rashtriya Swayamsevak Sangh has historically been and continues to be "to this day the backbone of Hindu nationalist organization."[75] All these groups share an ideological commitment to Hindutva.

Since the 1980s, but especially in the 1990s, Hindu nationalism has taken shape as a cultural movement that has gained legitimacy in Indian society, most notably among the Indian middle classes and elites, but, importantly, also across caste and class divides. Hindu nationalism has also emerged as a viable alternative to the secularism of the Congress as an adequate, even desirable, paradigm for national governance. In 1984 the Hindu nationalist BJP obtained a mere two seats in parliament following national elections. Since 1989, however, the party has twice formed the national government and once provided outside support to a national government. From 1998 to 2004, the BJP led the coalition NDA government, which served its full term until the next parliamentary elections were held in 2004.[76]

The broad context that provided the structural condition of possibility for the revival of Hindu nationalism was the failure, real and perceived, of the Congress Party to make good on its promises, despite the party having had the opportunity to rule India for most of the period from 1947 to 1996. While secularism and socialism were fundamental to the professed ideology of the party, in practice the Congress was inconsistent and cynical in its commitment to both, especially after Nehru's death in 1964. Commentators agree that the decay of the Indian political field set in with the rise to power of Indira Gandhi, who succeeded her father as prime minister. As seen in the previous chapter, scientific and technological development, while remaining central to the agenda of national progress, largely benefited the elites in postcolonial India. After a half century of Congress rule, neither poverty nor illiteracy had been significantly eradicated. The structures of distribution of income and wealth were by no means egalitarian. Caste-based discrimination and violence had not disappeared. And secularism had not succeeded in doing away with religious tensions. The Indian state had been regularly challenged by secessionist movements from Assam in the east to Punjab in the west and Kashmir in the north, as well as by Maoist movements in several states opposed to oppression perpetrated by the state and upper-caste landowners on tribal and peasant communities. Since independence, India had fought three wars with Pakistan and one with China. The Indian state, under Indira Gandhi's leadership, had played a central role in the formation of Bangladesh as an independent state, and, relatedly, in the breakup of Pakistan. Indian intervention in the ethnic conflict in Sri Lanka between its Tamil population and the Sinhala majority and state had resulted in Rajiv Gandhi's assassination. To say the least, India's relationship with her neighbors had been strained since independence.

One way to track the ascendancy of Hindu nationalism since the early 1980s is to link it to key events that have functioned as turning points around which the movement gained visibility and mobilized support among the Hindu communities of India. Five such events may briefly be described. The first turning point is the conversion to Islam by several hundred Dalits in Meenakshipuram, a village of Tamil Nadu, in 1981, as a protest against discrimination by the relatively higher caste Thevar community.[77] The event sparked off a strong reaction from the Vishwa Hindu Parishad (VHP), a social and cultural Hindu organization.[78] The reactionary discourse of the VHP emphasized the threat of India being overrun by Islam and Muslims through religious conversions carried out by force and funded by a global Islamic network, in particular, by the oil money of the Saudis and other Gulf countries. Since 1981, the VHP has also embarked on a long-term program of an aggressive counterconversion agenda, targeting both lower-caste and tribal communities as well as Muslims and Christians.

The second event is the Shah Bano case judgment of the Supreme Court of India in 1985 and the consequent actions of the Congress government to appease the conservative (and, to an extent, self-appointed) leadership of the Indian Muslim community. In 1985 the Supreme Court ruled that a divorced Muslim woman was entitled to a monthly maintenance under the Criminal Procedure Code.[79] The All India Muslim Personal Law Board, an intervener in the case, spearheaded a campaign against the judgment, portraying it as un-Islamic and an infringement on Muslims' right to live according to their own personal laws, and questioning the right of non-Muslim judges to interpret the Qur'an. The campaign was taken up across the country by conservative sections of Muslim male leaders. Parliament enacted a new law, which granted Muslim divorced women time-bound maintenance.[80] Hindu nationalist organizations described the actions of the Rajiv Gandhi government to the Shah Bano controversy as an example of the Congress tradition of appeasing Indian Muslims.

The third event is the destruction, in December 1992, of the Babri Masjid, a long-disputed religious structure claimed by Hindus and Muslims. Some Hindus say that the mosque, named after the Mughal emperor Babur, had been built by demolishing a temple that marked the birthplace of Lord Ram, the Hindu god. In 1984 the VHP started a campaign to "liberate" the mosque. On December 6, 1992, a Hindu nationalist-led mob destroyed the monument.[81] Hindu nationalist organizations and mobs were also responsible for targeting Muslims in bloody Hindu-Muslim riots in 1992 and 1993 that followed the demolition.

The fourth event is the anti-Muslim violence in Gujarat in the early part of 2002, which followed the burning of one train car of the Sabarmati Express, at Godhra on February 27, allegedly by a mob of Muslims.[82] The retaliatory violence was carried out with the support of the chief minister of the state, Narendra Modi, and with the apparatus of the state government. While the state-sponsored anti-Muslim pogroms in Gujarat may have cost the Hindu nationalist movement some support at the national level, other Hindu nationalist ideologues have lionized Narendra Modi as the true savior of Hindus.

Finally, the attacks on the World Trade Center of September 11, 2001, have also been used by the Hindu nationalist movement to further its agenda. Post-9/11 Hindu nationalist discourse attempts to locate Indian civilization as one of the first victims of an ongoing project of global Islamic domination and terror. The claim is that Hindu suffering at the hands of Islam has not been recognized in the international community nor given its due as a credible historical fact.

In the current moment, Hindu nationalism continues to target Indian religious minorities, often using the same kinds of arguments found in Savarkar's writings. Indian Muslims are labeled variously as outsiders, the descendants of Islamic invaders, unpatriotic, and pampered minorities. Indian Christians are, likewise, depicted as unfaithful converts who have betrayed their motherland by adopting a foreign faith. With a deep-seated antipathy to communism and socialism, Hindu nationalism describes the Indian Left as a coalition of anti-Hindu and antinational ideologues. It also targets Indian "pseudosecularism" as a doctrine of appeasement that unfairly privileges Indian religious minorities at the expense of the majority Hindu population.

Hindu nationalism does not restrict itself exclusively to any one arena of Indian social existence. It attempts to intervene in the arena of faith and religion directly and indirectly, it seeks to control state power, and it aims to influence the judicial system and to change the laws regarding minority rights. The movement has also challenged the depiction of historical events in school textbooks with the objective of "correcting" an alleged Leftist bias in these texts. During the period of NDA rule, the BJP government appointed several scholars with known Hindutva leanings and strong anti-Left views on important advisory and policy bodies, who then commissioned several revisions in school textbooks. The changes essentially amount to recasting the period of Mughal rule in India as an age of Islamic despotism, constructing an imagined tradition of seamless Hindu resistance and Hindu consciousness in opposition to the impact of Islam and Christianity, and glorifying the role of the Hindu nationalists in the Indian freedom struggle.[83]

In this manner, Hindu nationalist discourse attempts to reshape the cultural, political, social, legal, and educational fields in India. It seeks to intervene in every field in society, even as the political arm, the BJP, seeks to implement changes *across* fields through the agency of the state by the capture of political power. Commenting on the pervasive character of the development of Hindu nationalism in the public sphere in Indian society, Hansen observes that

> Hindu nationalism has emerged and taken shape neither in the political system as such nor in the religious field, but in the broader realm of what we may call public culture—the public space in which a society and its constituent individuals and communities imagine, represent, and recognize themselves through political discourse, commercial and cultural expressions, and representations of state and civic organizations.[84]

The unifying principle across these initiatives is the insistence, in each case, on Hindutva as a necessary component of Indian social life. For instance, since the 1980s, Hindu images started appearing with greater frequency in advertising, television series, and films; Hindu "values" gained greater prominence as an element of political discourse, not just in the rhetoric of the BJP but also in that of the Congress; and public events, even those organized by state institutions, began to be marked by Hindu rituals with increasing regularity.

Neoliberalism aims to reconfigure the nomos of the state by redescribing the social in terms of the economic. Through interventions in state and society, Hindu nationalism seeks to reconfigure the nomos of the Indian state by redescribing the social in terms of the cultural. As a foundational principle, Hindutva aims to replace the notion of secular citizenship with that of Hinduness. Acceptance of the fact that being Indian means being Hindu (as defined in Hindutva ideology) is proposed by Hindu nationalist discourse as the very condition of entry in *any* field, a principle meant to govern the practices of everyday life in Indian social space. As Bhargava puts it, the Hindu nationalist movement "now shapes the common sense of the very large Indian middle class. Its goals, beliefs, emotions, and prejudices are fast penetrating the discourse of ordinary people."[85] Herein lies the seduction of Hindu nationalism: in its appeal that it can display the core truths about Indian social reality, stripped of the obfuscations of socialism and secularism and cleaned of the ideological baggage of the Congress Party. The logic of Hindu nationalist discourse—that it is merely revealing the truth about the social—is strikingly resonant with the logic of neoliberal discourse. Both discourses essentially operate in the same manner: purporting to be *objective* truths about the world. The consonance between the two discourses produces a third overlapping discourse, technocultural Hindu nationalism, most visibly manifested in the sphere of technology.

TECHNOCULTURAL HINDU NATIONALISM

The doxas of both Hindu nationalism and neoliberalism posit the same general claim to revealing the Indian real, albeit from different angles—cultural and economic, respectively. Both discourses have been found persuasive by enough inhabitants in Indian social space to have permanently entered the structure of the state apparatus and various fields in Indian society. Technocultural Hindu nationalism emerges as a result of the convergence of the doxa of Hindu nationalism and of neoliberalism, by combining the cultural reason of Hindu nationalism with the economic and meritocratic logic of neoliberalism into a new nomos for the Indian nation-state.

Importantly, technocultural Hindu nationalism reflects a *particular* combination of Hindu nationalist and neoliberal ideologies. It accepts the claims of neoliberalism with regard to the economic sphere, but not with respect to other spheres. It does not necessarily accept notions of individual rights and freedoms on which neoliberal perspectives rest. Nor does it necessarily accept the neoliberal notion that all social behavior can be viewed in terms of a model of market relations in which free individual agents seek to maximize their gains. It combines the economistic logic of neoliberalism with claims about the cultural primacy of Hinduism; it accepts the neoliberal endorsement of free trade and open markets but ascribes the motivations of actors in the economic (and other realms) to a Hindu ethos, not to rational self-interest; and while it does not necessarily seek to abrogate Indian constitutional democracy, it privileges loyalty to the Hindu majority over any commitment to individual political or civil rights. In political terms, then, technocultural Hindu nationalism is a socially conservative interpretation of neoliberalism.[86]

The nomos of technocultural Hindu nationalism defines India as a culturally distinct and strong Hindu civilization that is simultaneously a technologically and economically powerful state in a globalized world. Even as it refashions neoliberalism within a cultural nationalist framework, Hindu nationalism has positioned itself as a necessary handmaiden to liberalization after 1991. Rajagopal observes that "the Hindu nationalist right thus came to prominence in India as the cultural complement to economic liberalization, which began to be instituted in the latter half of the decade."[87] Bhatt, similarly, contends that

> of significance from 1996 was the importance paid by the BJP to the radically different economic and political global environment and the rise of new information and scientific technologies to which India had to respond...A proud and confident Hindutva cultural nationalism (*sanskritik rashtriyavad*) emerges here both as a precise response to political, economic and social globalization and as the means for India's economic, political, technological and military rejuvenation on the world stage.[88]

There are several good reasons why the term *technocultural* is appropriate in describing this newly emergent variant of Hindu nationalism. For one, its adherents are drawn in good measure from communities of technology professionals who have been among the prime beneficiaries of India's liberalization and participation in a global economy. To be more specific, the proponents of Hindu technocultural nationalism are to a significant extent also the beneficiaries of India's post-1991 policies of IT- and Internet-oriented development and of the global demand for Indian technological expertise and software programmers. Indeed, the most significant overlap among the respective groups of supporters of Indian liberalization and Hindu nationalism is found among those who inhabit the field of technology. There is an emphasis on the quality of scientificity in the discourse of this new incarnation of Hindu nationalism in which both Hinduism and technology are depicted as embodiments, materializations, and representations of a deeper ethos of scientificity. And the use of the Internet—a symbol and embodiment of both a globalized world and an ascendant Indian nation—is central to the production and expression of technocultural Hindu nationalism.[89]

Profiling the Technocultural Hindu Nationalists

A profile of the main categories of supporters and beneficiaries of Hindu nationalism and Indian liberalization reveals how technocultural Hindu nationalism has emerged as a powerful force in Indian society. These groups can be categorized in mutually nonexclusive ways as Indian urban socioeconomic elites; relative subaltern groups within a broadly defined Indian elite; and a diasporic Indian Hindu population. It bears noting that not all the social groups that support Hindu nationalist organizations are uncritical advocates or supporters of liberalization. Hindu nationalism has also boosted its legions of supporters by exploiting the anxieties of those threatened by globalization.[90] Some groups in the Hindu nationalist family, such as the Swadeshi Jagran Manch, have also spoken out against globalization.

However, scholars agree that the core support for the Hindu nationalist movement, as is the case for liberalization, is drawn from India's "middle classes" or elites. Hansen argues that "Hindu nationalism could not consolidate any major constituency among the millions of marginalized and poor Indians. Instead, the xenophobic discourses of Hindu nationalism developed in the heart of the large and expanding middle class."[91] Support for the ideologies of Hindu nationalism and liberalization appears to have increased among Indian urban socioeconomic elites, who have switched loyalties from the Congress to the BJP. Vanaik locates urban elite support for Hindu nationalism and liberalization in the general conservatism of these groups.[92] He argues that these classes are attracted to the idea of a strong state, especially one that seems able to check the lawlessness represented by the messy mass politics of India's lower castes and rural populations.[93] Corbridge and Harriss view both Hindu nationalism and liberalization as "elite revolts."[94] The authors argue that while the agendas for each movement are not distinct and are sometimes contradictory, they are "often surprisingly complementary."[95] Both these movements are "vehicles for the interests and aspirations especially of the middle classes and upper castes."[96] Vanaik as well as Corbridge and Harriss also note that India's urban socioeconomic elites provided strong support for the BJP in the 1996 and 1999 general elections in India. As noted earlier, urban socioeconomic elites are also the main participants in the technological field—the stake setters and chief possessors of technological capital.

Internal changes within the broadly defined category of Indian elites (urban or rural) also explain other pockets of support for the Hindu nationalist movement. One bastion of support is comprised of the population of new entrants in the educational field in independent India. Mazumdar notes that

> with the rapid expansion of two-and four-year colleges as well as universities in the 1960s and 1970s, suddenly hundreds of thousands of students entered postsecondary educational structures, many from families that had not had access to university education before.[97]

By providing social services such as counseling as well as structures for cultural interaction and exchange, the Hindu nationalist movement could entrench itself among participants in the educational field. By 1983, the Akhil Bharatiya Vidyarthi Parishad (ABVP), a national student organization associated with the Hindu nationalist movement, "could count 250,000 members and 1,100 branch organizations" in universities across India.[98]

Gramscian analyses of contemporary Hindu nationalism suggest that the ideologies of socialism and secularism of the Congress Party have been found inadequate by many social groups in meeting their economic aspirations as well as their ideological-cultural demands.[99] The ideologies of the Congress, in other words, have not been accepted as the nomos of an economic or cultural field by numerous groups. Hindu nationalism has been successful precisely because it has managed to articulate itself as an overarching ideology that has resonated with diverse urban and rural groups. Hindu nationalism has especially proved to be a viable alternative to Congress secularism for the economic and ideological aspirations of intermediary urban and rural castes. As discussed earlier, Bose identifies two tiers of Indian elites.[100] The first tier consists of the highest caste and upper-middle-class communities of Brahmins and Kshatriyas that have monopolized bureaucratic and professional jobs in India since the colonial era.[101] The secondary category is broadly upper caste, but one notch below the first tier in terms of caste and lower-middle class in socioeconomic terms. Consisting of "highly urbanized castes of merchants, moneylenders, shopkeepers and small-scale entrepreneurs," this second tier has formed a traditional base of support for Hindu nationalism.[102]

It is difficult to establish whether these mercantile, largely lower-middle-class communities that have always supported Hindu nationalism also uniformly support liberalization. However, scholarly work suggests that many *new* entrants to the Indian middle class, whose very entry into the middle class has been enabled by liberalization, are strong supporters of Hindu nationalism. Bhargava argues that along with the decline of the Congress, the global crisis of socialism and secular humanism, and the growth in influence of the Hindu diaspora, a "fourth, and perhaps most important, reason" for the rise of Hindu nationalism is the emergence of new groups that have gained in economic and cultural capital.[103] He points out that "the success of Indian democracy and rapid changes in the economy in the neoliberal era brought wealth and enhanced social status to many hitherto marginalized urban and rural people."[104] The framework of secularism had "legitimated a social and political order that offered them few benefits," and, consequently, these groups were drawn to the antithetical ideology, Hindu nationalism.[105] These "new middle classes" are every bit as insecure as traditional elites about losing their privilege, and that anxiety finds expression in an antipathy to the policy initiatives of the Indian state for other disenfranchised communities, namely, "those at the bottom of the ladder."[106] Hindu nationalism, with its opposition to affirmative action for lower castes and its antiminority stance, appears as an attractive ideological allegiance for these groups. With the advent of liberalization, a wider section of Indians is availing of increased educational and professional opportunities. A proportion of these "new" middle classes might well have emerged as a result of their participation in the technological field, one of the main areas where economic liberalization has engendered professional opportunity

The Diaspora and Technocultural Hindu Nationalism

A key constituency of support for Hindu nationalism is an overseas, largely U.S.-based, Hindu population.[107] Consisting of NRIs and Persons of Indian Origin (PIOs), the diasporic Hindu community also includes technologically qualified professionals in the United States on H1-B visas and students pursuing higher education. Overseas Hindus provide strong financial support to Hindu nationalist initiatives in India. They are also active in promoting Hindu nationalist causes through numerous initiatives of a local, national, and global nature, from town-level community events in the United States to global conferences. Vinay Lal suggests that diasporic Hindus are drawn toward the narrow Hindutva model of identity out of an anxiety about their "invisibility" as a community in American life.[108] Rajagopal argues that Hindutva enables diasporic American Hindus to dispel casteist anxieties about being considered black in the United States.[109] Hindutva enables a compact between the homeland and the diasporic community, the exchange of economic capital for cultural assurance. Nostalgia for the homeland can now be enjoyed without the guilt that one has abandoned one's home. At the same time, the diasporic Indian can justifiably be critical of India as a concerned nationalist.

Mazumdar notes that most of the new Indian immigrants to the United States

> are not members of the older, established cosmopolitan Indian bourgeoisie. The vast majority come from small-town backgrounds, and their access to higher technical education in independent India changed family fortunes when they found jobs as doctors and engineers in the United States.[110]

While the possessors of considerable amounts of economic capital, the new groups lack the cultural capital of the old elites of the Nehruvian era, typified by the Oxbridge graduates who liberally peopled the Indian bureaucracy. Identification with the Hindu nationalist movement—in return for contributions, monetary or otherwise, to Hindutva causes—provides cultural capital in exchange of financial capital. As Mazumdar notes, the new groups are

> acutely resentful of the exclusionary politics of the older Indian metropolitan bourgeoisie whose class advantages placed them in better universities, their pride is assuaged when heads of major organizations and political parties from India come to visit them for the dollars they control.[111]

Many first-time graduates from Indian universities, who were socialized into Hindu nationalist ideology by the ABVP on Indian university campuses in the 1960s and 1970s, "constitute the new immigrants from India, and they include in their numbers the much touted thousands of computer and software experts."[112] It is thus not an exaggeration to conclude that the technocratic communities, who make up a sizeable proportion of the diasporic Indian Hindu population in the United States, are significantly populated by Hindu nationalists.

NRIs are viewed in post-liberalization India as typifying an Indianness worth emulating. Hindu nationalist discourse claims that it is a commitment to Hindutva that grants the Indian his or her real identity. Combining the two, the diasporic Hindu nationalist is now increasingly enshrined in the popular imagination as the perfect Indian-Hindu. The ideal vision of Indian society, correspondingly, is one where the cultural reason of Hindutva colludes with the economic reason of neoliberalism. The scientificity of Hindu culture is a key attribute in this reimagining of the Hindu-Indian nation.

The Emphasis on Scientificity

The authority of science and technology is used to affirm technocultural Hindu nationalism in mutually supportive and symbiotic ways. The attribute of scientificity is invoked as the basis of the truth value of claims made in support of the discourse. Relatedly, those who hold scientific and technological qualifications are granted—and arrogate for themselves—the authority to speak for the community and the nation. In this regard, Nanda observes that

> Hindu nationalists are obsessed with science…They use the vocabulary of science to claim that the most sacred texts of Hinduism—the Vedas, the Upanishads which contain the essence of Vedic teachings (also called the Vedanta) and Advaita Vedanta—are, in fact, scientific treatises, expressing in a uniquely Hindu idiom, the findings of modern physics, biology, mathematics, and nearly all other branches of modern science. From the most orthodox priests, to the modern globe-trotting, English-speaking swamis, academics (many of them in American and Canadian universities), pamphleteers and journalists, down to ordinary men and women sympathetic to Hindu nationalist ideology, all proclaim that Hinduism is simply another name for science.[113]

The seminal importance granted to science and technology can be explained by several factors. For a good number of the supporters of Hindu nationalism—whether in India or overseas, whether drawn from the ranks of old or new Indian elites—science and technology, especially since liberalization, have been the means for attaining educational, economic, and cultural capital. Science and technology are also described by the Hindu nationalists as egalitarian and meritocratic fields, when compared to the humanities or social sciences. Scientific ability is defined as a form of universally viable and universally appreciated talent that can help its possessors overcome any lack of educational, cultural, and economic capital.[114] Hindu nationalists argue that the study of history, literature, and the social sciences, especially in India, has been the preserve of the old Indian elites—the urban cosmopolitan, Westernized, Leftist cliques who are anti-Hindu.

The attribute of scientificity is, consequently, read into Hinduism, and, conversely, Hinduism or Hindu values are read into science and technology. Chakravartty observes that

> corporate networks between India and the United States promote a cyber-capitalist rereading of Hindu values, locating the success of high tech Indian entrepreneurs in essential characteristics associated with ethno-religious identity. For example, Indian competence in the Internet economy is associated with the "web of interrelations" that tie together Indian families across national borders.[115]

Along with scientificity, the attributes of globality and universality are also often emphasized in the discourse. There is a twofold rationale for this emphasis: the first reason is the self-evident value of both attributes in a globalized world, and the second reason is that the truth claims of Hinduism must be universally valid and globally applicable in order to qualify as scientific.

The contemporary Hindu nationalist emphasis on science and technology stands in a curious relationship to the Nehruvian vision of science and technology and to the anticolonial Indian nationalist vision of science and technology. On the one hand, the prestige accorded to science and technology in Hindu nationalist discourse is in continuation with the status granted to science and technological qualifications in both prior eras, the anticolonial nationalist and the Nehruvian. Yet the *basis* on which value is accorded to science in contemporary Hindu nationalist discourse is closer to the anticolonial nationalist rationale for valuing science and technology: the idea that ancient Indian (Hindu) civilization was a highly scientifically advanced civilization. In this reasoning, the Nehruvian emphasis on "scientificity" as a form of secular rationality is bypassed, as is the link between scientificity, development, and social justice. This critically important aspect of the nomos of the Indian state and the nomos of the Indian technological field during the Nehruvian era is rejected and abandoned, while the Hindu universalism projected on to science in the anticolonial nationalist era is refashioned, with a view to incorporating recent technological advances such as the Internet in the discourse. Scientificity is recuperated and recast as a Hindu cultural quality, as, indeed, is technicity. As in the Nehruvian era, however, technicity—that is, the accumulation of practical and professional skills delinked from a discourse of rationality and enlightenment—is affirmed as an authentic form and manifestation of scientificity.

The Internet and Technocultural Hindu Nationalism

Technocultural Hindu nationalism typically finds expression on the Internet. Two obvious reasons for this are, simply, expertise and access. Mazumdar observes that the U.S.-based Indian computer engineers and software programmers, who were socialized in Hindu nationalist organizations in their university days in India, "help maintain over five hundred VHP Web sites (www.vhp.org) with their messages of Hindutva, Hindu history, and Muslim-bashing."[116] Rajagopal points out that "with the proliferation of software engineers from India, the internet has become a site for expansion" of Hindu nationalist discourse.[117] Vinay Lal notes that culturally conservative Indian software programmers in the United States are the creators of many Hindu nationalist Web sites.[118] Unlike other media, such as newspapers and magazines, television, or books, participation in the virtual public arena of the Internet does not require apprenticeship in an industry or credentialization from institutions or industries. The Internet is perceived as a radically democratic space, free from the ideological manipulations of the Nehruvian secularists and socialists who, the Hindu nationalists contend, control the academy, the educational system, and the media in India. The Hindu nationalists view the Internet as a weapon through which they can counter the narratives of the Marxists who control the old Indian establishment. Rajagopal describes it accurately when he states that "the internet is seen not simply as a channel of communication, but as a key battleground that must be occupied and defended as the always copious varieties of 'pseudo-secular' criticism proliferate."[119]

Not surprisingly, the Web sites of "cyber-diasporic Hinduism" betray an obsession with proving their historical credibility and authority.[120] Abounding in terms such as "holocaust" and "genocide," Hindu nationalist Web sites simulate an academic language in describing the state of Hindu victimhood.[121] The sites attempt to produce a narrative that is historically authoritative and politically urgent: the truth about a historically beleaguered people that must be brought to light. However, this historical authority is sought to be located in a more fundamental scientific authority. On Hindu nationalist Web sites, for example, it is not uncommon to find articles where events in "Hindu history," such as from the *Ramayana* or *Mahabharata*, are claimed to be proved in accordance with the protocols of scientific research.[122]

The qualities of globality and universality are also highlighted on these Web sites. Hindu nationalist history is produced in the context of a spatial inversion—the globe is configured in the image of the nation. The Web sites promise that the global spread of Hindutva, the ideology of the Hindu nation, will subsume and reinscribe the world as a global Hindu nation-state. Cyber-diasporic Hinduism, in this fashion, often highlights relatively obscure elements of Hindu nationalist discourse, such as the VHP's "expansionist program for the Indian nation-state," to encompass the entire globe under the Hindu nation in the next century.[123] The global present is transcendentalized according to a logic of Hindu history; the simultaneous, homogeneous time of the imagined community of the nation-state is refashioned as the eternal time of the global Hindu nation.

The India of the twenty-first-century Hindu nationalists is founded on a forgetting of the achievements of the Nehruvian Indian state, no matter what its flaws might have been. The irony is that technocultural Hindu nationalism could not have emerged had it not been for Nehruvian nationalism and the Nehruvian emphasis on science and technology. Technocultural Hindu nationalism thus reveals a profound paradox: the progression of a partial democratization of society toward undemocratic ends, ideologies, or values. This book concludes with an examination of how the Internet may contribute to the complexities of online Hindu nationalist discourse. But before that, it is worth examining what is particular to the substance of Hindu nationalism online and the extent of the pervasiveness of themes such as scientificity and globality in the discourse. The next chapter, accordingly, presents a comparative analysis of the incidence and relevance of themes in online Hindu nationalism and other online Indian nationalisms.

ENDNOTES

1. Global Hindu Electronic Networks, *Hindu Universe*, (accessed September 4, 2008). This is the most recent date on which the site was accessed. The analysis and reading reflects the incarnation or version of the Web site as of February 19, 2005. The list of sections and linked sites presented here is meant to be representative—not exhaustive. As with many Web sites, some content on the site is accessible through multiple paths and is categorized under different heads.
2. More detailed close readings of Hindu nationalist Web sites are undertaken in Rohit Chopra, "The Virtual State of the Nation: Online Hindu Nationalism in Global Capitalist Modernity," in *South Asian Technospaces*, ed. Radhika Gajjala and Venkataramana Gajjala (New York: Peter Lang, 2008) 153–178, and Rohit Chopra, "Global Primordialities: Virtual Identity Politics in Online Hindutva and Online Dalit Discourse," *New Media and Society* 8, no. 2 (2006): 187–206.
3. Global Hindu Electronic Networks, "About This Site" (accessed February 19, 2005). When accessed on September 4, 2008, however, the page did not mention the relationship with the Hindu Students Council. For an explanation, see endnote 38, chapter 1.
4. Ibid., emphasis in original.
5. "About HSC" (accessed January 4, 2006).
6. Ibid.
7. "HSC Chapters," *Hindu Students Council*, n.d., http://www.hscnet.org/chapters.php (accessed January 4, 2006).
8. See Priya Ganpati and Ranjan Basu, "Pravasi Bharatiya Divas: Economy and the Diaspora," *rediff.com*, January 12, 2004, http://specials.rediff.com/money/2004/jan/10sld1.htm (accessed February 9, 2005).

   We can assume that the majority of NRIs and PIOs are Hindus, reflecting the demography of the Indian nation-state. The citizens of Pakistan and Bangladesh are not counted as NRIs or PIOs. For estimates of members of the Indian diaspora in relation to country, see Government of India, Ministry of External Affairs, Non Resident Indians & Persons of Indian Origin Division, "Estimated Size of Overseas Indian Community: Country-Wise," *Report of the High Level Committee on the Indian Diaspora*, December 2001, http://indiandiaspora.nic.in/diasporapdf/part1-est.pdf (accessed February 11, 2005). As per U.S. Census Bureau figures based on the latest count in 2000, there are 1.67 million Indian Americans in the country, of which 1 million were born in India. See Ela Dutt, "28.4 million foreign-born residents in U.S.," *Indian-American Center for Political Awareness*, n.d., http://www.iacfpa.org/p_news/nit/iacpa-archieve/misc/24052002-census-1.html (accessed February 17, 2005). As per the 2001 census, the Hindu population of India was approximately 827.6 million, which represented 80.5 percent of the total Indian population. Government of India, Office of the Registrar General & Census Commissioner, India, "Census and You—Religion," *Census of India*, n.d., http://www.censusindia.gov.in/Census_And_You/religion.aspx (accessed June 22, 2008).

9. As Vinay Lal points out, "The post-industrial civilization of North American Hindus is also, if a paradox may be entertained, a Vedic civilization." *The History of History*, 242.
10. *Hindugifts*, n.d., http://hindunet.tolshop.com/ (accessed February 11, 2005).
11. *Hindunet Signature Merchandise*, n.d., http://www.cafepress.com/hindunet2 (accessed February 11, 2005). The *Hindunet Signature Merchandise* Web site is an online shop located and hosted on the *Cafepress.com* network. *Cafepress.com* is an e-commerce service where anyone, individuals or companies, can set up an online shop. See *Cafepress.com*, n.d., http://www.cafepress.com/cp/info/ (accessed February 11, 2005).
12. *Mobilehindu*, n.d., http://www.mobilehindu.com (accessed February 11, 2005).
13. See also Global Hindu Electronic Networks, "Dharma & Philosophy," *Hindu Universe: Hindu Resource Center*, n.d., http://www.hindunet.org/dharma/ (accessed February 11, 2005).
14. Global Hindu Electronic Networks, "Interfaith Relations," *Hindu Universe: Hindu Resource Center*, n.d., http://www.hindunet.org/interfaith_relations/ (accessed February 12, 2005).
15. Global Hindu Electronic Networks, "Conversions" (accessed February 12, 2005).
16. Global Hindu Electronic Networks, "Relationship Between Hindu and Moslem" (accessed February 12, 2005).
17. Global Hindu Electronic Networks, "History" (accessed February 12, 2005).
18. See, for instance, Global Hindu Electronic Networks, "In the Islamic Ages" (accessed February 12, 2005). In an important and pioneering analysis of Hindu nationalist revisionist histories on the Internet, Vinay Lal describes the essentialist construction of Indian faiths, the presence of Hindutva ideology, and the antiminority sentiment on sites linked to the Global Hindu

Electronic Networks site and similar sites. See *The History of History*, 252–261.

19. See *The History of History*, 242, and Nanda, *Prophets Facing Backward*, 65. For a detailed analysis of the Hindu nationalist obsession with science, see especially 37–122.
20. Bourdieu analyzes neoliberalism in France and Europe in "The Myth of 'Globalization'" and in "Neo-liberalism, the Utopia (Becoming a Reality) of Unlimited Exploitation," in *Acts of Resistance: Against the Tyranny of the Market* (New York: The New Press, 1998), 94–105.
21. Bourdieu, "The Myth of 'Globalization,'" 31.
22. Bourdieu, "Neo-liberalism," 94.
23. Bourdieu, "The Myth of 'Globalization,'" 29.
24. Ibid., 30.
25. Ibid., 34.
26. Ibid., 37.
27. Ibid., 38.
28. Bourdieu, "Neo-liberalism," 102.
29. Bourdieu, "The Myth of 'Globalization,'" 33.
30. Ibid., 32.
31. Ibid., 34.
32. Ibid., 32, 34.
33. Anne O. Krueger and Sajjid Chinoy, "The Indian Economy in Global Context," in *Economic Policy Reforms and the Indian Economy*, ed. Anne O. Krueger (Chicago: University of Chicago Press, 2002), 21.
34. For example, the view that the necessity of liberalization and globalization must be recognized as simple facts prior to ideology is expressed in the *India Infrastructure Report 2002*, a document offering policy analysis and recommendations for the governance of commercial activity in India. Chiding the Indian state for not allowing the rupee to take a free fall against the dollar, one of the editors of the report proclaims that "this is not the time for ideological histrionics." Sebastian Morris and Rajiv Shekhar, eds., *India Infrastructure Report 2002: Governance Issues for Commercialization* (New Delhi: 3i Network and Oxford University Press, 2002), 1.
35. See, for instance, Muppidi, *The Politics of the Global*, 53.
36. Sachs et al., "Preface to the Paperback Edition," xi.
37. Greenspan, *India and the IT Revolution*, 109.
38. Wolcott and Goodman, "Global Diffusion of the Internet I," 565 (accessed May 17, 2005). The reasons provided by the authors for the centrality of the Internet to India's IT development are as follows: (a) the Internet is at the core of, and symbolizes, the very "notion of 'IT' in the late 20th and early

21st century" and, relatedly, Internet-related metrics signify the technological capability of any nation; (b) "the Internet itself is an enabling technology" that can transform commercial and personal interaction; (c) the Internet is a mass technology; and (d) the Internet is a "driving factor" spurring change in related "technologies, government policies, laws and services" (ibid., 566).

39. The other factors are "abundant supply of labor, low wages, cheap satellite communications." See Sachs et al., "Preface to the Paperback Edition," xi–xii.
40. Wolcott and Goodman, "Global Diffusion of the Internet I," 564 (accessed May 17, 2005).
41. Ibid., 609.
42. See Simone Cecchini and Monica Raina, "Warana: The Case of an Indian Rural Community Adopting ICT," April 2002, Information Technology in Developing Countries Working Paper, http://ssrn.com/abstract=568181 (accessed October 13, 2008).
43. Greenspan, *India and the IT Revolution*, 57.
44. Ibid., 58. For more on Indian technology professionals in Silicon Valley and the United States, see chap. 4, "Marginal Capitalisms," 55–76. See also chap. 3, "Rising Technopolises," in Arvind Singhal and Everett M. Rogers, eds., *India's Communication Revolution: From Bullock Carts to Cyber Marts* (New Delhi: Sage Publications, 2001), 130–182.
45. Chakravartty, "The Emigration of High-Skilled Indian Workers to the United States: Flexible Citizenship and India's Information Economy," 327. See also Stephen McDowell, *Globalization, Liberalization and Policy Change: A Political Economy of India's Communications Sector* (New York: St Martin's, 1997), cited in quotation.
46. Singhal and Rogers, *India's Communication Revolution*, 131.
47. Manjeet Kripalani and Mark. K. Clifford, "India Wired," *Business Week*, Asian edition, February 21, 2000, http://www.businessweek.com/2000/00_08/b3669008.htm (accessed June 17, 2003).
48. Thomas Friedman, "Globalization, Alive and Well," *New York Times*, September 22, 2002, sec. 4, 13.
49. Keniston, "Introduction," 17.
50. Ibid., 17–18.
51. Ibid., 18.
52. Ibid., 18.
53. Khilnani, *The Idea of India*, 148. This professional class has gradually been emerging since the 1970s in India, but "has been sustained and been given substantial economic power by the arrival in India, especially after the liberalization begun in 1991, of foreign capital and multinationals" (ibid.).

54. Ibid., 149.
55. Chakravartty, "The Emigration of High-Skilled Indian Workers to the United States," 326. Chakravartty takes the concept of "flexible citizenship" from Aihwa Ong, *Flexible Citizenship: The Cultural Logics of Transnationality* (Durham: Duke University Press, 1999). The relationship of expatriates such as workers in high-tech fields to their countries of origin can be described as "a flexible attitude toward citizenship," that is, "flexible practices, strategies, and disciplines associated with transnational capitalism that create new capitalist heroes of Asian modernity." Ong, *Flexible Citizenship*, 17–19; quoted in Chakravartty, "The Emigration of High-Skilled Indian Workers to the United States," 326.
56. Chakravartty, "The Emigration of High-Skilled Indian Workers to the United States," 327.
57. Ibid., 346.
58. Mupiddi, *The Politics of the Global*, 56.
59. Greenspan, *India and the IT Revolution*, 72.
60. Ibid., 108.
61. Keniston, "Introduction: The Four Digital Divides," 15.
62. "Number of Subscribers in Cellular (including WLL) and Fixed Line in India (2000 to 2005)" (accessed January 10, 2006).
63. "Number of Internet Subscribers in India (As on 13.03.1999 to 30.06.2005)" (accessed June 26, 2007).
64. "India: Internet Usage Stats and Telecommunications Market Report," (accessed June 19, 2007).
65. Wolcott and Goodman, "Global Diffusion of the Internet I," 608 (accessed May 17, 2005).
66. Ibid., 613.
67. Ibid., 609.
68. Ibid., 612.
69. Ibid., 613.
70. Ibid., 612. The report also cites an article that points out that "usage of non-English sites appears to be small. The Gartner Group estimates that only 11% of Indian users visited an Indian language site." See M. Rao, "India on Way to Digital Millennium, Madanmohan Rao Reports From the Gartner Summit in Mumbai, Aug. 14, 2000," *Inomy.com*, http://www.inomy.com/midd2.asp?storyid=1200134&sectioncode=17&subsectioncode=52&page=Subsection%20PageINOMY%20EXCLUSIVESE-COMMERCE; quoted in Wolcott and Goodman, "Global Diffusion of the Internet I," 612 (accessed May 17, 2005).

71. Wolcott and Goodman, "Global Diffusion of the Internet I," 613.
72. Ibid.
73. Ibid., 610.
74. Keniston, "Introduction: The Four Digital Divides," 15.
75. Khilnani, *The Idea of India*, 161.
76. In the 2004 national elections, the Congress emerged as a surprising winner. The Congress leads the current coalition United Progressive Alliance (UPA) government. The current prime minister of India is Manmohan Singh, who was finance minister in the 1991 Congress government that introduced economic liberalization, and is widely considered to be the architect of the 1991 reforms.
77. See Amberish K. Diwanji, "Why Do These Hindu Groups Always Seek to Turn India Into a Hindu Pakistan?" *rediff.com*, February 4, 1999, http://www.rediff.com/news/1999/feb/04akd.htm (accessed March 11, 2005).
78. The name of the organization roughly translates as the World Hindu Council.
79. See Nawaz B. Mody, "The Press in India: The Shah Bano Judgment and Its Aftermath," *Asian Survey* 27, no. 8 (August 1987): 935–953.
80. The new law was termed the Muslim Women (Protection of Rights on Divorce) Act 1986.
81. For an overview of the controversial history of the structure, see "Timeline: Ayodhya Crisis," *BBC News*, U.K. edition, last updated July 5, 2005, http://news.bbc.co.uk/1/hi/world/south_asia/1844930.stm (accessed August 15, 2005). See also A. G. Noorani, ed., *The Babri Masjid Question, 1528–2003: A Matter of National Honour*, vols. 1 and 2 (New Delhi: Tulika Books, 2003).
82. Dionne Bunsha, "The Facts from Godhra," *Frontline* 19, no. 15, July 20–August 2, 2002, http://www.frontlineonnet.com/fl1915/19150110.htm (accessed July 10, 2003).
83. See the section, "History Writing in India and Dangers From the Far Right," *South Asia Citizens Web*, http://sacw.insaf.net/India_History/index.html (accessed July 10, 2004).
84. Hansen, *The Saffron Wave*, 4.
85. Bhargava, "The Cultural Nationalism of the New Hindu," 11.
86. This is different from "social neoliberalism," or "forms of globalization more responsive to social needs and values," the kind of alternative framework for globalization that is presented at the World Social Forum annual gatherings. See Philip G. Cerny, Georg Menz, and Susanne

Soederberg, "Different Roads to Globalization: Neoliberalism, the Competition State, and Politics in a More Open World," in *Internalizing Globalization: The Rise of Neoliberalism and the Decline of National Varieties of Capitalism*, ed. Susanne Soederberg, Georg Menz, and Philip G. Cerny (New York: Palgrave Macmillan, 2005), 28.

87. Rajagopal, "Hindu Nationalism in the US," 470.
88. Bhatt, *Hindu Nationalism*, 174–175.
89. The gendered nature of the Indian nation and state in the technocultural Hindu nationalist imaginary warrants brief comment here. In technocultural Hindu nationalism, technology is affirmed as the means for a powerful India to take its rightful place on the global stage. The rhetoric of the strong Indian state in the age of globalization is in keeping with the masculinist construction of the Indian state in Savarkar's thought. Those who have possessed technological qualifications in India have historically been authorized to speak for the nation. By association, the acquisition of technological skills and qualifications is defined as a masculinist, patriotic task of protecting the honor of the Indian nation. Technocultural Hindu nationalism, like Hindu nationalism at large, is a gendered discourse, but the mode in which that gendering occurs changes with liberalization and globalization.
90. See Ghanshyam Shah, "Tenth Lok Sabha Elections: BJP's Victory in Gujarat," *Economic and Political Weekly*, December 21, 1991, 2921–2924.
91. Hansen, *India's Saffron Wave*, 7.
92. Vanaik, "Consumerism and the New Classes in India," 230–232.
93. Ibid.
94. Corbridge and Harriss, *Reinventing India*, xix.
95. Ibid.
96. Ibid.
97. Mazumdar, "The Politics of Religion and National Origin," 239.
98. Ibid. The Akhil Bharatiya Vidyarthi Parishad translates as "All India Student Association."
99. See Bose, " 'Hindu Nationalism' and the Crisis of the Indian State," 104–164; and, generally, Achin Vanaik, *The Painful Transition: Bourgeois Democracy in India* (London: Verso, 1990).
100. Bose, " 'Hindu Nationalism' and the Crisis of the Indian State," 132.
101. Ibid.
102. Ibid.
103. Bhargava, "The Cultural Nationalism of the New Hindu," 16.
104. Ibid.

105. Ibid.
106. Ibid.
107. See Das, "The End of Geography," 31–62; Vinay Lal, "The Politics of History on the Internet," 179–221; Rajagopal, "Hindu Nationalism in the US," 467–496.
108. Vinay Lal, "The Politics of History on the Internet," 189.
109. Rajagopal, *Politics After Television*, 251–252.
110. Mazumdar, "The Politics of Religion and National Origin," 247.
111. Ibid.
112. Ibid., 239–240.
113. Nanda, *Prophets Facing Backward*, 65. The text offers a detailed analysis of the Hindu nationalist obsession with science. *See esp*. 37–122.
114. There is little acknowledgment by technocultural Hindu nationalists that they are *also* relatively privileged, in comparison to Indian communities from the lowest rungs of the caste system, Dalits, and religious minorities. The strategic invocation of relative subaltern status holds particularly true for the new entrants among elites who are privileged in the context of India's population of more than a billion yet see themselves as victims of the old elites.
115. Chakravartty, "The Emigration of High-Skilled Indian Workers to the United States," 343.
116. Mazumdar, "The Politics of Religion and National Origin," 240.
117. Rajagopal, "Hindu Nationalism in the US," 484.
118. Vinay Lal, "The Politics of History on the Internet," 179.
119. Rajagopal, "Hindu Nationalism in the US," 485.
120. Vinay Lal, "The Politics of History on the Internet," 179.
121. Ibid.
122. See, for example, P. V. Vartak, "The Scientific Dating of the Mahabharat War," *Hindu Universe*, n.d., http://www.hindunet.org/hindu_history/ancient/mahabharat/mahab_vartak.html (accessed January 11, 2005).
123. See Vinay Lal, "The Politics of History on the Internet," 216, note 36; See also 204.

Chapter 6

Mapping Hindu Nationalism Online: A Comparison of Indian Nationalisms on the World Wide Web

This chapter undertakes a statistical mapping of Hindu nationalism, Sikh nationalism, and Kashmiri nationalism on the World Wide Web to identify the frequency and salience of relevant constituent themes in the online representations of the respective nationalisms.[1] Using the statistical technique of correspondence analysis (CA), the chapter maps the relative spread of such themes in relation to the three online Indian nationalisms. CA is a method that "analyzes the relations between the categories of two discrete variables" and enables the results of analysis to be presented in visual form.[2] For the analysis, variable *x* was defined as *Online Indian Nationalism*, and variable *y* was defined as *Salient Ideological Themes*. The categories of variable *x* were defined as Hindu

Nationalism, Sikh Nationalism, and Kashmiri Nationalism. The categories of variable *y* were defined as the following seven themes: Globality, Rights and Freedoms, Holocaust and Genocide, Terror and Violence, Scientificity, Universality, and Truth and Knowledge. Each theme was selected based on a case for its relevance to the general phenomenon of online Indian nationalism. Each theme was further conceptualized as comprising constituent subthemes; for example, the theme of scientificity included, among others, the subthemes of technology, chemistry, and mathematics. The primary hypothesis to be tested was that the theme of scientificity is most strongly associated with online Hindu nationalism in comparison to its association with the other online Indian nationalisms.

A sampling strategy was employed of scanning the World Wide Web through assessment of the top one hundred results of five keyword searches for each nationalism on the search engine Google in order to identify relevant resources for the particular nationalism.[3] On the basis of an assessment of the results for each keyword search, thirty-six unique resources were identified for Hindu nationalism, fifteen for Sikh nationalism, and twenty-one for Kashmiri nationalism. Frequencies for all relevant occurrences of subthemes and, accordingly, frequencies for all themes were counted using a concordance text analysis program. Frequencies of themes across the three nationalisms were weighted according to a base of $N = 100$. The data were analyzed using the CA function on the software program JMP.[4] The results of the data were obtained in numerical as well as visual form. The results confirmed the hypothesis of the strongest association of scientificity with Hindu nationalism, compared to the association of scientificity with other nationalisms. The chapter also offers an explanation, with reference to cultural, historical, and political contextual factors, of the relative associations of various themes with the three online nationalisms in light of their differences and similarities. The primary focus of the chapter is the characteristics of online Hindu nationalism, with the other two nationalisms assessed largely for comparative purposes. The first section of the chapter describes the rationale and framing of the hypothesis. The next section highlights some key theoretical and methodological issues for mapping the Web for online content analysis. The third section explains the construction and definition of the variables and the constituent categories of each variable. The fourth section explains the operationalization of the variables. The fifth section details the sampling technique and strategy. The sixth section offers a brief description of the method of CA. The seventh section presents the results in numerical tabulated and visual form. The final section offers an exploratory reading of the results.

Rationale and Framing of the Hypothesis

As described earlier, Hindu nationalism takes on sociological specificities in cyberspace, which derive from historical characteristics intrinsic to the discourse as well as from larger structural factors, such as the impact of globalization and technological development on Indian society in the 1990s. Both these constitutive dimensions—internal and external—combine in complex ways in online Hindu nationalism. However, the extent to which the characteristics of online Hindu nationalism derive from its embeddedness in the economies of Internet production, exchange, and circulation merits deeper investigation. The question is whether the prominent manifestation of scientificity in online Hindu nationalism is *primarily* a function of the online, virtual status of the discourse, as opposed to a central, substantive ideological feature that is both amplified and reformulated online. Although scholars have noted the importance of the theme of scientificity in Hindu nationalist ideology, the inquiry is apposite given the communication possibilities enabled by the Internet and given the profile of technology professionals as among the chief producers and consumers of Hindu nationalist ideology online. Put another way, the issue is whether the characteristics of online Hindu nationalism might be more reflective of the *medium* of the Internet than of the *content* of Hindu nationalist ideology itself. A second, related, question concerns the empirical manifestation of the dominance and visibility of the theme of scientificity in online Hindu nationalism. If—as argued earlier and also suggested by others—the quality of scientificity is defined in Hindu nationalist discourse as a core characteristic of

Hindu faith, culture, and identity, it follows that the theme of scientificity should appear prominently in online Hindu nationalism across Web sites, portals, discussion groups, and the like that are dedicated or sympathetic to the ideology. In other words, the incidence of the theme of scientificity in online Hindu nationalism should be significantly large in any mapping or sample of the discourse online.

A comparative analysis between Hindu nationalism and other Indian nationalisms with regard to the relative distribution of themes within and across the nationalisms has been chosen as a suitable framework for answering these questions. The argument motivating the comparative analysis can be stated as follows: if the emphasis on scientificity in online Hindu nationalism is a function primarily of its virtual, online status, then other comparable online Indian nationalisms should *also* indicate more or less the same degree of occurrence of the theme. If, in contrast, scientificity is an ideological characteristic *particular* to online Hindu nationalism, it should be most strongly associated with Hindu nationalism in comparison to its association with other Indian nationalisms. The other, complementary, proposition tested is that scientificity is among the most salient themes in online Hindu nationalism in terms of the respective frequencies of occurrence of various themes (although testing this second proposition does not necessarily require comparison with other nationalisms). The hypothesis, framework of analysis, and methodology have been designed such that these propositions may be effectively and accurately tested.

MAPPING THE WEB: SOME THEORETICAL AND METHODOLOGICAL ISSUES

An assessment of critical issues from five select studies suggests some strategies for adequately conceptualizing representations of discourse in cyberspace. Singh and Baack's analysis of American and local Mexican corporate Web sites as culturally specific communication and marketing resources and tools conceptualizes the Web as a "cultural document."[5] Using "cultural values" as an analytical unit, the authors identified four values—collectivism, high power distance, high uncertainty avoidance, and masculinity—which were then operationalized.[6] For example, collectivism, as reflected in Web-based communication, was operationalized in terms of indicators of community relations and community-oriented sentiment. These indicators included themes of family and nation, loyalty programs on Web sites, and avenues of community interaction such as chat rooms.

For American companies, the authors drew a sample of Web sites of American companies that sold products in both locations, America and Mexico, from the Fortune Global 500 list on the magazine's Web site (www.fortune.com). There were 30 Web sites that had "fully developed and usable American and Mexican" pages.[7] For local Mexican sites, out of 103 sites listed on the Wright Investors' Service site (www.wisi.com), the top 35 companies were selected. For these 95 Web sites, twenty to twenty-five pages from each site were analyzed, amounting to a total of about two thousand pages of Web content. Out of eight hypotheses regarding the variation in elements of cultural specificity, four were strongly or partially confirmed, one was marginally confirmed, while three were not supported. For example, it was "clear that family theme, tradition theme, use of local terminology, proper titles, and rank or prestige of the company" were areas in which American and local Mexican Web pages showed differences irrespective of whether the target audience was Mexican or American.[8] Mexican Web sites appear to emphasize collectivism and local terminology as well as "a clear depiction of gender roles" with men chiefly shown as performing company functions.[9]

Tateo's study takes as its object of analysis the online network of Italian extreme right-wing organizations.[10] The study employs a social network analysis model combined with content analysis and conceptualizes a set of Italian extremist Web sites as "a community based on common interests and mutual legitimation."[11] The study classifies Web sites as "extremist" based on evidence of occurrence of certain themes, specifically, White Supremacy, Overt Racism, Symbolic Racism, Negationism (denial of events such as the Holocaust), and Conspiracy Theory. For the study, the corpus of sites was taken from a Web directory, Portale della destra in Italia (Portal of the Right Wing in Italy). Eleven categories of Web sites were identified: Music groups, labels, fanzines, and music portals; Italian Social Republic (ISR) nostalgics; Skinhead groups; Revisionists and negationists; Cultural, traditional Catholics, and new age and neo-mystical groups; Neo-nazis; Publishers, reviews, and commercial sites selling books or merchandizing; political parties or movements; Militaria; National revolutionaries, nationalitarians, and communists; and Others. The study analyzed a network of seventy-seven sites, divided into these eleven categories in terms of their linkages and relations. The results revealed the existence of a "coherent network" among these various far-right sites, with four sites from different categories occupying the most prominent position among the network at large. The results "might represent what we could call an 'offer of differentiated Internet content' by the ER, which exists as a loosely structured and interacting galaxy of groups under the label of Extreme Right."[12]

In a review of research on personal home pages, Doring asserts that Web pages represent a "new object of sociological, psychological, linguistic and communication studies research."[13] The author defines a personal or private home page on the Web as one "published and maintained by an individual who may or may not be affiliated with a large institution."[14] The defining characteristic of such a page appears to be its contrast with pages published by "organizations, institutions or formal groups," such as corporate or nonprofit sites.[15] The definition derives from the status of the Web page owner or producer and not so much from the nature of content, since even an individual Web page concerned with professional activities qualifies as a personal home page. At the same time, the personal Web page need not be necessarily created by an individual owner himself or herself. It may be created by Web designers, a couple, family, or a small group. For the study, the author sampled Web pages from three sources: Web directories, Web rings, and link lists. Doring's analysis of sampled pages centers on questions such as the characteristics of home page owners; the extent to which home pages are regularly maintained by owners; issues of form and content; the perceived audience of the Web page; motivations for creation of home pages; and visibility of gender effects in construction of pages. The author also offers a classification of personal home pages by different types of criteria. In conclusion, Doring notes the heterogeneity of home pages, the diversity of home page producers and consumers, and the complexity of processes of home page production and consumption.

Holt presents an analysis of two contrasting sets of political Web pages and Web sites as representations of political ideology.[16] The first set of Web pages and Web sites consists of five official candidate Web sites in the 2000 U.S. presidential campaign, comprising three versions of the site of the Republican candidate, George W. Bush, and two versions of the site of the Democratic candidate, Al Gore. The second set of pages is from a single activist Web site. Holt employs a dual strategy and method of reading these resources. On the one hand, Internet communication can be viewed "conforming to a linear or transmission (primarily monologic) model of communication."[17] On the other hand, the dynamic nature of Internet communication indicates that producers and consumers of Web site content are engaged in a dialogic conversation that is open-ended as well as inclusive and reflective of multiple voices.[18]

Marschall's analysis of ideological discourses of "community" on career-related sites focuses on specialized "third-party Internet career sites," specifically, Monster.com, Jobs.com, and Guru.com, as well as the recruitment sections of corporate Web sites.[19] Analyzing discourses of community on the sites, Marschall draws attention to the tropes of "heroic figures," "cultural icons," and "mythic belief structures" as well as a distinctive use of language on the sites. The heroic figure is an idealized construction targeted at job applicants, suggesting a model candidate who would be best suited to gain from the services offered on the sites. Cultural icons refer to symbols, such as merchandise offered for sale, through which the Web sites aim to establish a bond with and among site users. Mythic belief structures—overarching frameworks that guide commonsensical perceptions—are reflected in narratives romanticizing American capitalist virtues such as "the self-made man," the "triumphant entrepreneur," "rugged individualism," and the glorification of technology.[20] Marschall's analysis reveals that both a discourse of community and a discourse of power are at work in these Web sites, which are salient in view of the instability wrought upon the old model of the American labor market by changes such as technological development. However, discourses of community and power are configured differently in each type of site. While Internet career sites effectively operated as a "personal empowerment domain," the recruitment sections of corporate Web sites were presented as a "corporate opportunity domain," where individual success was linked to successful incorporation within a larger corporate identity.[21]

Some general insights can be drawn from the studies that are helpful for our purposes here. Generally, Web sites or Web resources can be taken as representative of ideological discourses. As Holt notes, "Web pages/sites can be designed to express ideology and they can be read to discover circumstances that led to them being composed as they have been."[22] More specifically, Web sites may be classified as reflective of a particular ideology, culture, or worldview on the basis of key themes. While the preceding studies use different methods of analysis and sampling, all of them define the object of inquiry in terms of themes. Parsing Web content into themes appears to be an effective and useful strategy in determining units of cyber-discourse for analysis. An additional important point to note is that there is no fixed formula for sampling the World Wide Web. However, an effective strategy seems to be to use existing tools and resources such as Web directories or Web site lists, especially from reliable sources such as Fortune.com. In this regard, a search engine, especially a credible one such as Google, may provide an effective means of searching the Web. A final observation here is that there is no one single theoretical framework and methodology for sampling and analyzing Web content, even Web content of a particular type. For example, while there are several perspectives on analyzing content on political Web sites, "few methods can be used across instances of Web pages/sites."[23] Holt also observes that "there has yet to be formulated an overall conceptual vocabulary of Web pages and sites that will permit observers to draw conclusions based on comparing them to each other."[24] While the lack of such a vocabulary poses challenges, it also allows for some flexibility in conceptualizing online discourse in terms of a model most suited to the project at hand. As the studies indicate, the frameworks and particular methods chosen for analysis appear closely related to the study objectives. These broad considerations have been taken into account in defining the variable constructs, in operationalizing variables, and in choosing a sampling strategy for the analysis undertaken here.

Variable and Variable Category Constructs

This section describes the conceptualization and construction of the variables Online Indian Nationalism (variable *x*) and Salient Ideological Themes (variable *y*). The operationalization of the variables in terms of constituent categories is also explained. For variable *x*, the categories are Hindu Nationalism, Sikh Nationalism, and Kashmiri Nationalism. The conceptualization of variable *x* also covers a description of each nationalist movement. For variable *y*, the categories are Globality, Rights and Freedoms, Holocaust and Genocide, Terror and Violence, Scientificity, Universality, and Truth and Knowledge.

Construct of Variable *x*, Online Indian Nationalism, and Its Constituent Categories

The understanding of nationalism employed in this chapter draws on Deol's "working definition" of the nation as

> an imagined political community seeking a historic homeland for its own people, whose solidarity is sustained by the presence of cultural channels of communication, notably ethnicity and language and whose boundaries are in continual flux. In the modern world, any body of people who consider themselves to be a nation can claim the right to self-determination, or the right to a separate sovereign independent state for their territory.[25]

Nationalism can be articulated as the proposition that a majority ethnic, religious, or cultural group or community possesses the primary and exclusive right to inhabit and possess a territory as well as the right to legitimize that ownership through the institutions and framework of the modern nation-state. In addition to Hindu nationalism, two ongoing ethnic or cultural nationalist movements for an independent nation-state in India that meet this criteria—Sikh nationalism and Kashmiri nationalism—have been selected for analysis here. As the two most politically significant secessionist movements in the history of postcolonial India that combine religious and territorial logic, they are well suited for comparative analysis.[26]

Contemporary Hindu nationalism is a majoritarian cultural nationalist movement based on the ideological proposition that India is essentially a Hindu nation and should reflect the interests of the Hindu majority.[27] Hindu nationalist ideology has increasingly found support among a wider cross section of the Indian populace, especially the elite middle and upper-middle classes, in the last two decades. Although the BJP-led coalition was defeated in the last general election held in India in 2004, Hindu nationalism has taken shape as a strong force in Indian political life, the public domain, and social discourse at large.

In contrast, Sikh nationalism and Kashmiri nationalism are minority nationalist movements, each of which aims to establish an independent state in territory currently under Indian sovereignty. The Sikh nationalist movement seeks to establish the independent state of Khalistan in the northwestern Indian state of Punjab.[28] Defining the Sikhs primarily as a religious-cultural group, the Khalistan movement articulates its goal of independence in religious and political terms and can also be viewed as a religious fundamentalist movement. The roots of the Khalistan movement are to be found in the colonial period, when the British colonial government defined Indian communities in essentialist terms as religious collectivities. From a longer historical perspective, the Sikh nationalist cause can be located in terms of "the transition of the Sikhs from a religious congregation in the sixteenth century into an ethnic community in the eighteenth century and from an ethnic community into a nation in the late twentieth century."[29] However, it was only in the 1940s, when the prospect of the Partition of India and the creation of

Pakistan was raised by the Muslim League as a precondition for Indian independence, that "the Sikhs with a tragic desperation began to visualize the Punjab as their homeland."[30] With Partition, nearly two-thirds of the British province of Punjab went to Pakistan, while the rest went to India.[31] After independence, the province of Punjab was divided into two administrative units. In addition to the state of Punjab, another administrative territory known as the Patiala and East Punjab States (PEPSU) was created, where Sikhs and Hindus were nearly equal in number at 49.3 percent and 48.8 percent.[32] The Akali Dal, a Sikh organization, however, began an agitation for a Punjabi-speaking state or Punjabi *suba.* Deol argues that while the agitation was presented by the Akali Dal as a matter of linguistic identity, "the fundamental issue" was one "of the rights and claims of a minority community."[33] In 1956 PEPSU and Punjab were merged, but it was in 1966, with the reorganization of Punjab on linguistic grounds, that the Akali Dal achieved its objective. In 1966 the state was divided into Hindi-speaking Haryana and Punjabi-speaking Punjab. A part of Punjab was also added to a third Indian state, Himachal Pradesh. While in 1951 61 percent of Punjab was Hindu and 35 percent Sikh, with the reorganization in 1966, Hindus accounted for 44 percent and Sikhs almost for 54 percent of the population of Punjab.[34]

The demand for Khalistan is in part an outcome of this longer history of an emphasis on the distinctiveness of Sikh linguistic, ethnic, and religious identity, and in part an outcome of the complications of Indian politics. In 1977 the Congress had been defeated in elections in India. The Akali Dal had joined hands with the Janata Party as part of the coalition government that took power.[35] Toward reestablishing its presence in Punjab, the Congress cultivated and supported Jarnail Singh Bhindrawale, a Sikh preacher, using him to fracture the Akali Dal.[36] But in the bargain, the Indian prime minister, Mrs. Gandhi, created a Frankenstein. When the Congress returned to power in the 1980 elections and began to treat Bhindrawale as expendable, he took up the cause of Khalistan.[37] Khilnani notes that by this time Bhindrawale had gathered the support of a "small but energetic constituency of young men in Punjab (as well as in London's Southall and Acton)."[38] The early 1980s appear to have been a crucial turning point in the Khalistan movement. A speech by Ganga Singh Dillon in 1981 asserted that the Sikhs were a nation.[39] The Sikh secessionist movement was at its peak in the 1980s, with Sikh militancy rampant in Punjab and elsewhere in India. In 1984 the then prime minister, Indira Gandhi, gave the order for the Indian army to storm the Golden Temple, the holiest site of Sikhism, in the city of Amritsar in Punjab, on the grounds that the temple was harboring militants. In October 1984, Mrs. Gandhi was assassinated by her Sikh bodyguards in retaliation for her perceived desecration of the Sikh temple. By the 1990s, the Khalistan movement had effectively been quelled by the Indian state, but the demand for an independent Sikh state continues to be articulated by sections of the Sikh community, especially those located overseas.[40]

The Kashmiri nationalist movement seeks to establish an independent homeland in the territory of Kashmir that spans India, Pakistan, and China.[41] Kashmir has been a disputed territory between India and Pakistan since the inception of both states as independent political entities, reflecting the particular history by which Kashmir acceded to the Indian state in 1947. Since independence, India and Pakistan have fought three wars, the first in 1947 itself, the second in 1965, and the third in 1971. In 1999 Indian and Pakistani troops were engaged in a conflict in the Kargil region, with the potential for the conflict to turn into a full-fledged war. The border tensions between the states, however, were defused, and since that time, there have been no hostilities in the region of the magnitude of a political crisis.

Both India and Pakistan claim Kashmir in its entirety, though currently part of Kashmir falls under India, a part—known as Azad (or "free" or "independent") Kashmir—is controlled by Pakistan, while China also controls a segment of the territory. The Indian Constitution, however, grants a special status to the Indian state of Jammu and Kashmir as per article 370. The conflict is additionally complicated by the demand by segments of the Kashmiri population for complete independence, that is, affiliation with neither India nor Pakistan. Starting in the late 1980s and continuing through the 1990s, the Kashmiri secessionist movement has manifested itself in the form of militancy and terrorism, spearheaded by organizations such as the Jammu and Kashmir Liberation Front and Hizbul Mujahideen, as well as by relatively more recently formed groups such as Lashkar-e-Toiba and Jaish-e-Mohammad. The Kashmir struggle is defined through an often uneasy combination of religious and political logic. Pakistan's claim to Kashmir rests on the fact that at independence, as is the case now, the majority of Kashmiris were Muslim. Currently, the Indian state of Jammu and Kashmir is India's only Muslim-majority state, with 64 percent of the population identifying as Muslim as per the 1981 census.[42] According to the logic by which India was partitioned, Muslim-majority states would form part of Pakistan, while Hindu-majority states would belong to India. Although the Hindu ruler of Jammu and Kashmir, Maharaja Hari Singh, had signed an instrument of accession to India in October 1947, the Pakistani viewpoint has consistently been that this decision did not reflect the will of the Muslim majority of Kashmir and violated the right of Kashmiri Muslims to self-determination. Many of the secessionist outfits, especially those that have taken to militancy, have defined their struggle against the Indian state in terms of a religious framework, often as an Islamic *jihad.* In December 2001, the Indian parliament came under a terrorist attack, for which India holds Jaish-e-Mohammad and Lashkar-e-Toiba responsible.[43]

The movement for Kashmiri independence appears to be more robust than the Khalistan movement, given that, to a degree, the former has been internationalized as a global political matter. But the Khalistan movement has not died out entirely and has gained some visibility in the domain of the domestic politics of various countries such as the United States and the United Kingdom, as evidenced by resolutions in the U.S. Congress and the entry of Khalistan in the Unrepresented Nations and Peoples Organization (UNPO) at the Hague in the Netherlands.[44] While the specificities of each nationalist movement, as well as the differences between majoritarian nationalism and minority nationalism, are, no doubt, immensely significant, the three nationalist movements may be treated as manifestations of the same phenomenon on the basis of certain structural characteristics. The main common feature is the fundamental argument that a cultural, ethnic, or religious group possesses the right to inhabit and politically control its *own* nation-state. This argument is itself comprised of two propositions: the first is that a people can be defined, in essentialized terms, on the basis of cultural, ethnic, or religious identity; and the second is that a cultural, ethnic, or religious ethos, defined as the singular quality of a people, can be perfectly and ideally represented in the institutions of the modern, territorial nation-state. In the case of all three nationalisms—Hindu, Sikh, and Kashmiri—cultural or ethnic sovereignty is equated with the right to political sovereignty in the form of a state. There is also an additional general structural feature shared by the nationalisms, which is that none of the three movements have been successful in their stated ideological aims. Although the Hindu nationalist movement does not seek to constitutionally abrogate Indian democracy, and while it has achieved a significant measure of visibility in numerous arenas of public life, it has not succeeded in its ambition to redefine India as an incontestably Hindu nation or as a de facto religious state. Similarly, the Khalistan movement has not succeeded in separating Punjab as the Sikh homeland from the Indian nation-state, nor has Kashmir gained independence from India or Pakistan. As unrealized projects in this sense, all three nationalist movements may be described as ongoing struggles. This indicates a motivation on the part of the movements to *continue* to promote their respective ideologies in public social, political, and cultural forums. Had any of these movements succeeded in realizing their respective goals, their status as *nationalist* causes would have largely been a historical fact, and the chief reason for propagating the nationalist cause would no longer be extant. The nationalisms, in this hypothetical situation, would have been rearticulated as official ideologies of state, reflecting a somewhat different set of salient characteristics such as those pertaining to specific functions of state institutions.

In light of the preceding discussion, variable x in this chapter pertains specifically to the form taken by nationalist movements as they manifest themselves on the World Wide Web or cyberspace. Variable x, termed *Online Indian Nationalism*, refers to the ideological content of nationalism as it takes shape in cyberspace, subject to any possible reconfiguration through the modes of representation and expression enabled by the

Internet and, equally, reflecting the interests and composition of online producers and consumers of the ideology. The construct of Online Indian Nationalism reflects the understanding that the substantive ideological content of Indian nationalism will express itself in the medium of cyberspace in a particular form that overlaps but is not necessarily identical to "offline" expressions of Indian nationalism. Hindu, Sikh, and Kashmiri nationalism are the three constituent categories that fulfill the fundamental requirements of qualifying as nationalisms.

Construct of Variable *y*, Salient Ideological Themes, and Its Constituent Categories

The construct of variable *y*, or *Salient Ideological Themes*, is defined as a set of key themes that reflect both the substantive concerns of the three Indian nationalisms and the online status of these movements. The argument here is that a nationalist ideology can be parsed in terms of fundamental elements or principles that distil conceptually discrete, even if interlinked, claims that constitute the discourse. Examples of such elements might include the notion of unity between state and religion; the axiomatic claim that a nation-state should have an official religion; the claim to distinctness of the cultural identity in question; emphasis on the historical struggle of a people toward nationhood; claims and evidence of subjugation or oppression at the hands of another cultural group or the nation-state; and so on. Such constituent thematic elements or principles may reflect both substantive commitments as well as instrumental imperatives. Typically, some elements would be foregrounded as a tactical or strategic matter in particular contexts to gain visibility or to mobilize support from various constituencies. Additionally, in cyberspace, certain other elements, relatively irrelevant in an "offline" milieu, may take on additional importance. For example, given that the Internet is a transnational technological medium, nationalism in its online manifestation might well privilege an idiom and vocabulary of international law and human rights over the specificities of local issues and domestic struggles. An important point to note here is that the construct of variable *y* does not purport to be comprehensive or exhaustive; it does not seek to cover all crucial themes across the three nationalisms, Hindu, Sikh, and Kashmiri. The objective in the construction of variable *y* is threefold: (a) to include the *critically important* themes of Hindu nationalist discourse, such as scientificity, given the focus of the analysis; (b) to identify *some* common important themes across the nationalisms at large; and (c) to identify common important themes across the *online* incarnations of the nationalisms. The criteria reflect the primary focus on Hindu nationalism, with the other two nationalisms serving a comparative analytical purpose.

The selection of themes based on the aforementioned criteria led to an initial identification of five themes, as per the following rationale. The theme of Globality was identified on the basis of the argument that nationalisms would market themselves in cyberspace as *global* issues—of relevance to the international community at large as well as of relevance for diasporic ethnic, cultural, or religious groups dispersed across the world—to speak to the potentially global audience using the Internet. Nationalisms, as manifested online, would differ in this regard from their articulations in alternate circuits of transmission, such as print, meant primarily for consumption within the limits of the territorial nation-state. The theme of Rights and Freedoms was included based on the reasoning that a nationalist movement would typically emphasize the political right to sovereignty as well as other political, cultural, social, and economic rights, with reference both to domestic constitutional and legal frameworks and to the framework of international human rights, as defined, for example, in the UN Declaration of Human Rights.

Relatedly, it was assumed that a nationalist movement might seek to bolster its case for independence by drawing attention to a violation of the rights, freedoms, and liberties of individuals, communities, and political or cultural organizations committed to the cause at the hands of state and nonstate agents and actors. The theme of Holocaust and Genocide was included as a specific manifestation of the global emphasis of online nationalism. Preliminary research on the World Wide Web revealed that various Indian social and cultural groups sometimes described themselves online as the victims of a holocaust or genocide perpetrated either by another community or by the Indian state, or both. Some Hindu nationalist Web sites, for instance, speak of the genocide of Hindus over a millennium at the hands of Islamic invaders. There are at least two Web sites expressly dedicated to the Hindu holocaust, the *Hindu Holocaust Museum* and the *Online Hindu Holocaust Museum*.[45] Web sites that take up the cause of the Kashmiri Pandits, a Hindu community of Kashmir that has been displaced in large numbers from Kashmir as a result of violence at the hands of Islamic militant groups, describe their persecution as genocide.[46] Some Sikh sites speak of the holocaust and genocide of Sikhs in the anti-Sikh riots of 1984 that followed the assassination of Mrs. Gandhi. Web sites sympathetic to the cause of Kashmiri independence also speak of the holocaust perpetrated on the Kashmiris by the Indian state and Indian security forces. In light of these observations, the theme of Holocaust and Genocide was included as a category under variable *y* to test whether the phenomenon of the invocation of holocaust and genocide was uniformly pervasive across the nationalist discourses or varied significantly across the three online nationalisms. The theme of Terror and Violence, encompassing actions perpetrated by the state and nonstate actors, was included based on the following reasoning. Situations where the state is locked in militant conflict with secessionist movements are often characterized by human rights violations committed by the state as well as by secessionist groups. Members and supporters of the nationalist movement, as well as civilians, might be the victims of state violence and brutality. As the literature and Web resources on the Kashmir and Punjab conflicts unearthed at the sampling stage indicated, the human rights violations perpetrated by the Indian army were a key thematic concern. There is also the possibility that civilians are often the victims of the actions of militant secessionist groups. Indeed, opponents of secessionist movements point to the terroristic actions of such groups that result in the loss of life of innocent victims, in addition to extracting other social and economic costs. The fifth theme, Scientificity, was included, given its centrality to the hypothesis—that is, for the purpose of testing the claim that an emphasis on scientificity is particular and unique to the discourse of Hindu nationalism. Additionally, when the data collected through the sampling process were assessed, two other themes stood out in relative prominence in Hindu nationalism as compared to the other nationalisms. These two additional themes are Truth and Knowledge and Universality. As earlier chapters indicate, the Hindu nationalist proposition that Hinduism is a universal religion or way of life embodying objective, transcendental truths is interwoven with, and related to, the claim that Hinduism is a scientific faith. The two additional themes were, therefore, also included as other unique characteristics of Hindu nationalism.

Operationalization of Variables

This section explains the logic underlying the operationalization of variables for the purpose of analysis. The operationalization of variables was closely related to the sampling strategy and methodology of CA. The objective of the operationalization was to conceptualize the respective universes of variable *x* and variable *y* in empirical terms, such that *units* for each variable (and all constituent categories) could be accurately defined, identified, searched, and counted as part of a well-defined sample conducive to analysis through the method of CA. For variable *x*, Online Indian Nationalism, it was critically important to define online discourse in terms of discrete, identifiable, and enumerable units. Similarly, for variable *y*, Salient Ideological Themes, the key question was finding adequate, accurate, and countable indicators of themes.

Operationalization of Variable *x*

Two important concerns needed to be addressed in any operationalization of variable *x*. First, the domain of online discourse is open-ended, vast (even if finite) at any given point of time, and dynamic, as material is constantly being added, updated, or removed in cyberspace. The universe of online discourse—and by extension, of online nationalism—is not bounded, unlike, say, the universe of content contained in a set of five newspapers over a particular time period. Second, online discourse is expressed in numerous forms that, while overlapping in characteristics, cannot be viewed as variations of the same basic communication paradigm. For example, online communication and representation includes interactive paradigms such as instant messaging, chat rooms, discussion forums, and blogs, where content is highly dynamic; relatively more static sources of information such as Web pages, where content may be updated anywhere from daily to annually; archives or records of data; commercial gateways; multipurpose portals; dedicated radio channels; and the like. The entity of the "Web site" may be constituted in terms of any combination of these varied forms of communication and representation. To define the World Wide Web as made up of Web sites, while not inaccurate, is not very helpful as an empirical description of precisely what it is that is being collected or analyzed. Third, an object such as nationalist ideology or nationalism might manifest itself in diffuse fashion in cyberspace at large, concentrated and clearly identifiable in some Web sites or interlinked Web rings, while more nuanced and submerged in other forms of online representation. In light of these characteristics of online discourse, variable x was operationalized as consisting of *primary Web resources* rather than the more nebulous and fuzzy term *Web site*. The category of primary Web resource was also defined such that it would account for the inconsistency and fuzziness of content in cyberspace. More precisely, a primary Web resource, whether for Hindu, Sikh, or Kashmiri nationalism, was defined as comprising four important features. Each of these features is now briefly described.

First, a primary Web resource was defined broadly as a resource reflective of or sympathetic to a nationalism, Hindu, Sikh, or Kashmiri. The category of the primary Web resource draws on a distinction between "first-order" discourse, taken as evidence of the voice or self-presentation of the nationalism, and "second-order" commentary on the discourse. For example, some academic or journalistic writings on Hindu, Sikh, or Kashmiri nationalism that aim to comment on the phenomenon from a standpoint external to the discourse qualify as second-order commentary.

Even when such commentary may appear to agree with some aspects of a particular nationalist movement, the primary motivation for the piece is not the propagation of the ideology but rather the professional task of academic analysis or journalistic comment. It is true that the line between first-order discourse and second-order discourse is sometimes fuzzy. For example, Hindu nationalist Web sites often affect a pose of academic credibility. Similarly, there are some ideologues who engage online with scholarly debates, with the implicit or explicit claim that their analyses are marked by academic rigor.[47] Yet such resources are clearly sympathetic to, and can even be seen as part of, the Hindu nationalist cause.

Locating the relationship of a particular Web resource to a nationalist cause can sometimes be a difficult task, complicating the question of whether to define it as a primary Web resource or not. As noted earlier, for each nationalism, five hundred search results were assessed, consisting of the top one hundred results for each of the five search phrases. The sample for each nationalism was constructed from the set of five hundred search results. For the three nationalism samples cumulatively, fifteen hundred search results were assessed. At the sampling stage, borderline cases were assessed on a case-by-case basis. Several sections of the resource in question were evaluated in detail for additional indicators, and supplementary research was conducted toward determining the nature of the resource. A similar case-by-case assessment was carried out with regard to Web resources of organizations or individuals where there was no categorical statement signifying affiliation with, or support for, the nationalist ideology. Here, the answer was often provided by indicators such as links to other Hindu nationalist Web sites. In the case of Hindu student organizations in American universities, the Web resource indicated the formal affiliation of the organization with a larger Hindu network whose Web site, in turn, revealed its ideological affinity with the Hindutva cause. In other cases, a nationalist resource sought to reproduce an article or document originally published elsewhere that supported the nationalist claim or was linked to such supporting documents. Where the resource appeared largely comprised of external links or reproduced material, even if these were sympathetic to a nationalist cause, it was not included as a primary Web resource. An example of such a skeletal resource is the Web site *Islam/Muslims in Kashmir*, which is mostly made up of reproductions of links to and articles from other sites.[48]

Second, a primary Web resource was defined to cover resources containing *relatively* static and stable content on the Web, but not resources containing content generated through interactive, real-time paradigms such as instant messaging or live chat. The underlying rationale behind omitting such real-time content was straightforward: such conversations are typically private, whether restricted to two or more interlocutors. The objective was to identify resources meant for public consumption, where the nationalism is actively articulated as part of public discourse toward establishing its credibility, propagating its ideology, and mobilizing support for its cause. Private content was precluded from the ambit of the primary Web resource for reasons of relevance in addition to reasons of confidentiality. Also, while some chat rooms are public and chat sessions may be archived online, it is often difficult to identify a clear focus in such conversations. Importantly, the category of the primary Web discourse did include discussion forums, as distinguished from public chat rooms. Discussion forums are often dedicated to a particular topic, such as Hindutva or Hindu nationalism, and the ideological or thematic focus of the forum can easily be identified. Content generated in discussion forums is relatively stable and static. In discussion forums or groups, content is generated on specific topics and participants/members post responses by email. Conversations are documented, archived, and accessible as "threads." A "thread" consists of all responses to a particular topic and reflects the order in which successive email responses were received to the original topic posting.

The criteria of relatively stable and static content was used to identify primary Web resources reflective of the respective nationalisms across various types and forms of online representations such as personal Web pages, historical Web sites, Web sites of political and cultural organizations, online news magazines dedicated to a nationalist cause, government and nongovernmental Web sites, and so on. For example, a multipurpose portal—even if it included chat forums and commercial activity—did qualify as a primary Web resource as long as it contained relatively stable and static content and a clear articulation of the nationalist ideology. A stand-alone discussion forum categorically dedicated to the Hindutva cause was also included as a primary Web resource. However, on the basis of the same criteria, a stand-alone Hindu nationalist or Sikh nationalist chat forum was not included as a primary Web resource.

Third, a primary Web resource is a time-bound representation of online discourse. As explained in the section on sampling, the World Wide Web was mined for primary Web resources of each nationalism using the Google search engine and technology. The primary Web resources for each nationalism were sought to be identified within a limited time window of four days at most, given the possibility of content on any resource changing significantly over time as well as the possibility of changes in search rankings. Google displays ranked search results based on many factors, including relevance and number of visitors to a site. Realistically speaking, it was unlikely that, in the course of a few weeks, organizations, individuals, or groups represented on primary Web resources would significantly revise their ideological commitment or remove text such as a "motto" or site description reflective of that ideological commitment. It was also unlikely that over a couple of months, rankings in search results would change dramatically such that an entirely different set of results would show up through the Google search. However, even if remote, this possibility had to be taken into account, warranting the need to conduct the search in a short time period of a few days. At the same time, it was also possible that a Web resource might become inactive within a few days because of technical reasons such as server failure or possession of a Web domain lapsing due to nonpayment. A sample containing many inactive resources would be compromised in terms of its adequacy as representative of a nationalist discourse. It was, hence, equally important to ensure that the majority, if not all, of identified primary Web resources were still valid over the course of at least a few weeks. Given these factors, a strategy was adopted of defining the primary resources as time-bound representations of nationalisms, yet of enduring significance. Following this imperative, the searches for each online nationalism were conducted within four days, and unique primary resources for each nationalism were identified on the basis of the search. For any particular keyword, however, assessment of all one hundred search results was conducted on the same day. All primary Web resources were revisited about three weeks to a month later to check if the links were valid, and data from the appropriate sections or Web resource units were downloaded within two to four days of the second visit to the resources. The information collected through the data "freeze" in this manner enabled ease and accuracy of analysis. This also allowed the primary Web resources to be checked and rechecked accurately for distribution of themes (variable *y*) that might have been affected through any changes in the content over time.

Finally, and importantly, the primary Web resource was not defined as necessarily identical to the particular search result page that showed up through the Google search. Instead, the search result was understood to refer to the resource in its entirety. A relevant resource in its entirety was treated as the primary Web resource, with the key indicator taken as the "root" page. The root pages, as well as other sections of the resource, were visited to determine the location of the clearest and most categorical articulations of the ideology or ideological commitments on the Web sites. Typically, but not always, this would be found on the home page of the resource or on a section page such as "About Us," "Who We Are," "Our Mission," or the like. The Web page offering the clearest statement of ideology was downloaded as representative of the primary Web resource. It should be noted that in some cases, the home page itself showed up in the search results. Also, in other cases, on evaluation, the original-search result page itself was eventually found to offer the clearest articulation of the nationalist ideology. But the main point to be noted is that a Web resource as operationalized here is not necessarily limited or identical to a particular page displayed in search results. Rather, the page is understood as a part of a relevant Web resource, and it is the Web resource that is being sampled, not the specific page.

A hypothetical example will make this clear. Let us assume that a search through one of the keywords for Sikh nationalism displays an article on the Web site *Khalistan Nation* with the URL http://www.khalistannation.com. This article, titled "article 1," is located in the "Articles" section of the site. The location and URL of article 1 is http://www.khalistannation/article/article1.htm. In this case, the primary Web resource is not article 1 but the Web site *Khalistan Nation in its entirety*, the root URL for which is http://www.khalistannation.com. The home page, as well as the other sections, might indicate that the ideology of the site is most clearly presented on the "About Us" page with the URL http://www.khalistannation.com/about.htm, in which case this page ("About Us") would be selected as representative of the discourse. Article 1, the page actually displayed in the search, would not be selected as representative of the discourse.

Where the resource was hosted on a public domain such as Yahoo Geocities, the first page of the resource, not the home page of the site host, was taken as the root page. For example, one actual search result in one of the keyword searches for Hindu nationalism was the page with the URL http://www.geocities.com/jairama1/posters.htm. This page offers a collection of Hindutva or Hindu nationalist posters. It is a section of a Hindu nationalist Web site created by an individual and hosted on the Geocities service. The root page or home page of this resource, then, is taken as http://www.geocities.com/jairama1/ and *not* http://www.geocities.com (which is the home page of the Yahoo Geocities service).[49] In certain other cases, identifying the root page was somewhat complicated by the fact that the first page of the Web resource was merely a visual introduction in Flash technology or a skeletal umbrella structure offering little more than links. Here, on a case-by-case basis, another page that offered a clear picture of the focus and structure of the resource has been taken as the root page.

Operationalization of Variable *y* and Its Constituent Categories

For variable *y*, Salient Ideological Themes, the categories are the following themes: Globality (G), Rights and Freedoms (RF), Holocaust and Genocide (HG), Terror and Violence (TV), Scientificity (S), Universality (U), and Truth and Knowledge (TK). Each of these themes was further defined in terms of a fixed set of keywords. The sample of content collected for each nationalism would be checked for the number of occurrences of each keyword. The total frequency of all keywords in each set (corresponding to a theme) would be counted in each nationalism sample. Importantly, each individual occurrence of each keyword was also assessed with regard to the context of its use in a sentence. The objective underlying such assessment was to confirm whether the keyword was used in the specific and well-defined sense of the theme. To minimize error, counting of occurrences of keywords, as well the listing of all keywords by context, was not undertaken manually, but through a concordance program, the ConcApp Concordance and Word Profiler Version 4.[50] The respective keywords for each theme are as follows.

- For the theme of Globality, the keywords are Global, Globalization/Globalisation, Globally, Globalism, Globe, International, Internationally, World, Worlds, World-wide/Worldwide.
- For Rights and Freedoms, the keywords are Right, Rights and -rights (as in human-rights), Free, Freed, Freely, Freedom, Freedoms, Liberty, Liberation, Self-determination, Sovereignty, Sovereign, Democracy, Democratic, Independence, Independent, Violation, Violations, Violating, and Violated.
- For the theme of Holocaust and Genocide, the keywords are Holocaust and Genocide.
- For the theme of Terror and Violence, the keywords are Terror, Terrorism, Terrorist, Terrorists, Terrorized, Violence, Violent, Non-violence, Non-violent, War, Wars, Warfare, Kill, Killed, Killing, Killings, Death, Deaths, Prison, Prisoners, Jail, Jailed, Torture, Tortures, Tortured, Torturing, Riot, Riots, Fundamentalism, Fundamentalist, and Fundamentalists.
- For the theme of Scientificity, the keywords are Science, Sciences, Scientists, Scientific, Technology, Technologies, Technological, Technique, Medicine, Medical, Astronomy, Physics, Physiology, Physicians, Chemistry, Chemical, Engineering, Engineers, Nuclear, and Mathematics.

- For the theme of Universality, the keywords are Universal, Universally, and Universality.
- And, finally, for the theme of Truth and Knowledge, the keywords are Truth, Truths, Knowledge, and Wisdom.

It should be noted that the list of keywords was finalized through an iterative process, on the basis of assessment of the final samples of text collated for each nationalism. Some initially listed keywords (often plural forms or adjectives of relevant nouns) did not show up in the samples. For example, the words "Globalized/Globalised," "Internationality," and "Democracies" did not appear in any of the samples. The use of some keywords was simply too varied and diverse to qualify as corresponding to any theme. Some other relevant words for particular themes were suggested by the listing of a frequency count of *all* words in the respective samples. These listings are enabled by the concordance program. The goal in finalizing the set of keywords corresponding to each theme has been to obtain a reasonably comprehensive spread of words that can capture the incidence of a theme in a nationalist discourse at large. Misspelled instances of keywords have not been included in the frequency count of individual keywords and cumulative count of keywords for themes. The incidence of misspelled words appeared to be small in number, given that the units of variable *x* downloaded from the primary Web resource are largely written in correct English with few typographical errors. It is reasonable to assume that the number of misspelled words that have been ignored do not significantly affect the composition of the nationalisms with regard to the relative incidence of themes.

SAMPLING STRATEGY, PROCEDURE, AND SAMPLE DEFINITION

To identify relevant primary Web resources for each online nationalism—Hindu, Sikh, and Kashmiri—scanning the World Wide Web through a search engine was chosen as the most appropriate sampling procedure.

For this purpose, the search engine Google was selected as the most appropriate and effective search tool. Google has developed a proprietary technology known as PageRank that assesses over a hundred factors from 8 billion URLs on the World Wide Web, toward objectively measuring the importance of Web pages, which are displayed by rank based on the search term or phrase employed.[51] Google does not provide details about its technology and all the factors that determine a page ranking. However, the criteria used by the search technology in assessing a resource and ranking it in search results include the relevance of content as well as the extent to which a page is linked to other sites. The criteria used in Google's ranking system evaluate Web page and Web site relevance in terms of content, usability, and popularity.

For each nationalism, the objective of the search was to identify the widest possible range of relevant primary Web resources. It was, accordingly, decided to conduct multiple searches for each nationalism on the basis of general important keywords that would apply to the nationalism at large, across its varied manifestations. For each nationalism, five keywords were selected that reflect features of the nationalism at large as well as fundamentally important characteristics and key historical events pertaining to the nationalism. For each keyword, the top one hundred search results were individually assessed to see whether the resource displayed qualified as a primary Web resource or not. For each nationalism, five hundred search results have been individually visited and assessed, and for all three nationalisms cumulatively, fifteen hundred search results have been assessed.

The rationale for each keyword choice for each nationalism and the time frame for searches for each nationalism are detailed next, but, at the outset, one important aspect of the logic of search keyword selection needs to be explained here. The search keywords *do not* coincide with the terms or phrases designating the constituent categories of variable y, or Salient Ideological Themes, although, of course, it is hoped that the search results should reflect these salient themes. The reason for not using the particular category terms of the salient ideological themes as the search terms can be described as follows. It is extremely plausible that out of the universe of each online nationalism, some resources might emphasize one or more of the salient ideological themes more than other resources. In such a case, the search results might selectively be skewed toward displaying a higher concentration of those resources from the universe of the online nationalism at large. Using a more general term as a basis for the search would, in contrast, allow for a relatively more objective mapping of the nationalism, without risk of such bias. Given the assumption that each nationalism is comprised of salient ideological themes that may be found across the nationalism in cyberspace at large, a search using general terms should, in any case, identify Web resources where evidence of the salient ideological themes is abundant. An example will clarify this argument. Out of the universe of Hindu nationalist Web resources, it is plausible that some Web resources would deal extensively or even exclusively with the issue of the scientificity of Hinduism, while other resources would address the matter to varying extents. Searching the Web using the keyword "scientificity" would run the risk of prioritizing those sites that focus solely or largely on scientificity, in opposition to other sites that may also refer to scientificity but less extensively. Also, such a search might well display many pages from the *same* Web resource instead of displaying a more even spread of Hindu nationalist sites, all or most of which would also include matter on scientificity, but not necessarily with the same degree of emphasis. However, a general search for "Hindu culture" or "Hindu nation" would display a broader range of nationalist sites, which would reflect the core Hindu nationalist proposition about Hindu culture as the basis of the Indian nation. If—as is the claim made in this chapter—the theme of scientificity is widely spread across Hindu nationalist discourse, it follows that such sites displayed through a general search for "Hindu culture" or "Hindu nation" should anyway include content on scientificity.

For online Hindu nationalism, the following five keywords or phrases were identified as search strings for the searches, which were conducted between April 5, 2005, and April 8, 2005. The search strings were

(1) Hindutva, (2) Hindu Nation, (3) Bharat Mata, (4) Hindu Culture, and (5) Ram Janmabhoomi. Terms 2 and 4 are self-explanatory, with regard to their relevance for Hindu nationalist discourse. *Hindutva* is the term used to describe the ideology of the Hindu nationalist movement and literally means "Hinduness." Bharat Mata literally means "Mother India" and is often used in Hindu nationalist discourse to describe the Hindu homeland or the Indian motherland. Ram Janmabhoomi literally translates as "Birthplace of Ram." The phrase *Ram Janmabhoomi* refers to the site of the disputed Babri mosque in Ayodhya, which, Hindu nationalists claim, is the birthplace of Lord Ram, the Hindu deity. The liberation of the birthplace of Lord Ram has been a longstanding goal of the Hindu nationalist movement. In December 1992, a Hindu nationalist mob destroyed the mosque, an action that resulted in Hindu-Muslim riots in various parts of the country in 1992 and 1993. Building the Ram temple at the disputed site continues to be an objective of the Hindu nationalist movement.

For online Sikh nationalism, the searches were conducted between April 7, 2005, and April 8, 2005. The search strings employed were (1) Khalistan, (2) Sikh Nation, (3) Khalsa (4) Sikh Culture, and (5) Operation Bluestar. Khalistan is the name of the proposed independent Sikh homeland. The term *Khalsa* signifies purity and is an integral component of Sikh religious and cultural discourse. Operation Bluestar refers to the military operation conducted by the Indian state in 1984 to flush out secessionist Sikh militants from the Golden Temple, the holiest site of the Sikhs. In retaliation for the operation, the Sikh bodyguards of Indian Prime Minister Indira Gandhi assassinated her in 1984, an event that was followed by anti-Sikh riots in New Delhi in which thousands of Sikhs were massacred. The invasion of the temple further spurred the Sikh secessionist movement, and in Sikh nationalist discourse, the event has become a symbol of the authoritarianism and anti-Sikh bias of Indian state policies.

For online Kashmiri nationalism, the searches were conducted between April 11, 2005, and April 12, 2005. The search strings employed were: (1) Kashmir, (2) Kashmiri Nation, (3) Kashmiri Struggle, (4) Kashmir *Jihad*, and (5) Kashmiri Culture. Search strings 1–3 and 5 are self-explanatory. The term *Kashmir Jihad* was included since secessionist organizations in Kashmir often mobilize the rhetoric and ideology of *jihad*—that is, "holy war" or "holy struggle"—to describe their political struggle for independence.

The criteria for identifying and selecting a primary Web resource have already been spelled out. Some additional criteria were also used to identify the primary Web resources for particular online nationalisms, given the historical particularities and sociological specificities of each nationalist movement. The searches for identifying primary Web resources of Hindu nationalism did not specially target what may be termed "sub-Hindu" or regional Hindu nationalisms, exemplified by the ideology, policies, and actions of the regional Hindu nationalist party Shiv Sena or by Web sites of the Hindu Kashmiri Pandit population that appear to be in agreement with some aspects of Hindu nationalist ideology. However, such resources were also displayed in the results following the searches using the general keywords. They have been included in the final list of identified resources, wherever they qualified as primary Web resources as per the definition offered previously.

For Kashmiri nationalism, the primary focus was on identifying those resources aligned with movements seeking to liberate Kashmir from Indian control. The searches did not specially target other nationalist movements in the region such as the Balwaristan or Balochistan movements that also seek an independent state in the South Asian region, part of the territory whose proposed state may or may not overlap with Kashmir. However, where such resources were displayed in the search results, they have been included if they met the criteria of primary Web resources. One resource that was displayed in a search, the Mughalstan Web site, appears to be primarily an online movement. Indeed, there does not appear to be an "offline" Mughalstan political movement. However, since the Mughalstan Web site qualified as a primary Web resource, it has been included in the final list of primary Web resources for online Kashmiri nationalism. The keyword searches for online Kashmiri nationalism also displayed some Web sites reflecting the cause of the Hindu Kashmiri Pandit community. While these Web sites can also be viewed as Kashmiri and "nationalist," they are usually reflective of a Hindu majoritarian nationalism and opposed to the secessionist demand of separatist Kashmiri organizations. Kashmiri Pandit sites were thus not considered as representative of the Kashmiri secessionist movement. Finally, in the searches for online Kashmiri nationalism, pan-Islamic Web resources were often displayed in the search results. These were included as primary Web resources for online Kashmiri nationalism if there was clear evidence of sympathy for the Kashmiri nationalist cause and if there was a significant amount of content about the Kashmir issue. Such resources were assessed on a case-by-case basis, through visiting numerous pages and conducting internal searches on the resource to ascertain the extent, nature, and relevance of content on Kashmir contained in the site.

The distinction employed between a "first-order" primary Web resource and "second-order" commentary in the assessment of each search result has also already been described. As noted, the distinction was often not clear, and such resources had to be assessed on a case-by-case basis. In some cases, a second-order article originally published in a magazine or newspaper was reproduced on a nationalist resource. Here, the root URL of the nationalist resource was visited to determine whether the resource qualified as a primary Web resource. Similarly, affiliation of a particular Web resource to a nationalism, where not clearly identifiable, has been assessed on a case-by-case basis. For example, the Web site of the World Sikh Council, an umbrella body comprised of representatives of various Sikh organizations in the United States, does not include categorical evidence of affiliation with or sympathy for the Sikh nationalist cause.[52] The *World Sikh Council* Web site is also critical of inaction on the part of the Indian government in providing justice to victims of the 1984 anti-Sikh riots. The Web site also contains a link to the *Council of Khalistan*, an avowedly Sikh nationalist Web site.[53] But these links do not qualify as enough evidence to establish the *World Sikh Council* as a primary Web resource of Sikh nationalism. In this case, the ideological affiliation of the *World Sikh Council* to the cause of Sikh nationalism cannot be conclusively established, and the Web site has not been included as a primary Web resource.

Based on the top one hundred results for the five keyword searches for each nationalism, a final list of primary Web resources was drawn up for each nationalism, with the corresponding root URLs and site title or descriptor identified for each resource. Repeats of primary Web resources within and across keyword search results were not counted, since the goal was to obtain as wide a spread as possible of online nationalist resources. Not counting repeats across results for a nationalism, thirty-six unique resources were identified for Hindu nationalism, fifteen for Sikh nationalism, and twenty-one for Kashmiri nationalism. Initially, sixteen resources were identified for Sikh nationalism. However, one resource was inactive as of April 13, 2005, remained inactive as of May 15, 2005, and was accordingly discarded. For Kashmiri nationalism, initially twenty-two resources were identified. However, one resource was inactive as of April 28, 2005, and also as of May 15, 2005, and was thus discarded. Appendices A, B, and C provide the final list of unique primary resources for each online nationalism, including the root URL and title or descriptor for each resource.

The primary Web resources were revisited on May 5, 2005. Different sections of each Web resource were assessed to identify the best "unit" to select for the purpose of analysis. The appropriate page, identical to or containing the unit of text reflective of the nationalism, was downloaded to freeze the data. Any pages overlooked or not saved correctly on May 5, 2005, were downloaded again within the next ten days. Text was selected from any appropriate page or section of the Web resource, including but not limited to the home page and "About Us" section, wherever the clearest available articulation of the ideology and self-definition of the nationalism was available. Text was cut and pasted from each downloaded Web page into separate text files for each primary resource. In some instances, the pages did not save properly to the computer. In these instances, the pages were revisited within four days and text was cut from the online page and pasted into the text file.

The amount of text selected as representative of the unique primary resource—and constituting the unit of nationalism—varied depending on the particular resource. In most cases, the final text selected included the entire content of a page; however, in a few cases, it was a section or part of a page or a combination of two sections (for example, introductory text of a few lines describing the site focus and a "motto" that might be spelled out later in the page). The text selected did not include commercial advertisements, links, or other extraneous and irrelevant content. Also, certain pages included relevant descriptive text written on image files in gif and jpeg format that could not be cut and pasted into text files. Such relevant text found on images was manually entered into the text files. While at the stage of finalizing the lists of unique primary resources, repeated resources were not included, some units of text were found to be identical with others in the compilation of final text units. The reason for this is that several primary Web resources (which are unique and distinct from each other as Web entities in terms of their URLs) are part of a network of organizations broadly devoted to the same ideology. Content is shared and replicated across such networked sites, and it is possible that the same individual or groups of individuals may be responsible for developing content on these different sites. Also, several organization Web sites link to umbrella organization Web sites. Several sites may contain links to the same page in the umbrella organization Web site, where the ideology is most clearly spelled out. The "About Us" sections on several Hindu nationalist sites, for example, link directly to the "About Us" section page on another portal, *Hindu Universe*, where the affiliation and goal of the organization is presented.[54] The text on this page has been used as the representative unit for several unique primary Web resources of online Hindu nationalism.

Appendix D presents the list of particular Web pages, sections, and aspects—and combinations thereof—from which the units of text reflective of the ideology of the respective nationalisms have been constructed. Based on the exercise just detailed, thirty-six text files were prepared from the primary Web resources of Hindu nationalism, representing thirty-six units of online Hindu nationalism; fifteen text files were prepared from the primary Web resources of Sikh nationalism, representing fifteen units of online Sikh nationalism; and twenty-one text files were prepared from the primary Web resources of Kashmiri nationalism, representing twenty-one units of online Kashmiri nationalism. The set of files for each nationalism was further collated in three master files in text format, one each for Hindu nationalism, Sikh nationalism, and Kashmiri nationalism. Each master text file represented the final sample for the respective online nationalism.

THE METHODOLOGY OF CORRESPONDENCE ANALYSIS

This section very briefly describes some important aspects of the methodology of CA. CA is "basically an exploratory and descriptive" statistical technique that reveals associations in contingency tables.[55] The method of CA, therefore, does not test causality between two variables but rather correspondence between variables tabulated in rows and columns. The correspondence between rows and tables is presented in visual form, enabling further interpretation. Greenacre notes that "the primary goal of correspondence analysis is to transform a table of numerical information into a graphical display, facilitating the interpretation of such information."[56] While popular in France, CA is relatively less well known in the United States.[57] Of particular note is Bourdieu's use of CA in his text *Distinction*, where the methodology was used to test the association between the social profile of individuals and their cultural tastes, habits, or dispositions.[58]

The chief concepts of CA are "geometric ones rather than statistical ones."[59] Greenacre defines the three main concepts of CA as profiles, masses, and chi-squared distances.[60] Profiles refer to the relative frequency counts of variables by row and column. The profiles can also be understood as mathematical vectors that "define points in a multidimensional space."[61] Masses are the weights assigned to profiles, proportionate to the number that determines the profile.[62] The chi-square statistic is a measure of the discrepancy between expected and observed frequencies, alternately termed the measure of independence or homogeneity. As Greenacre notes, "[T]he larger this value, the more discrepant the observed and expected frequencies are, i.e., the less convinced we are that the assumption of homogeneity is true."[63] The inertia of a contingency table is the chi-square value relative to the total of the table.[64] Singular values indicate "the relative importance of the dimensions or how large a share of the total inertia each of them explains."[65] These indicators can be seen in the results of the data analysis presented next. The JMP Professional software (version 5.0.1.2) has been used for creating the contingency table, performing the CA, and producing the results in visual form.

Analysis of Results

Analysis of the data collected in the samples was undertaken as a two-step process. In the first step, the frequency for each theme for each nationalism was correctly counted, recorded, and weighted. In the second step, the tabulated data was entered into the JMP software program for the CA to be performed. The first step of the process was conducted through the ConcApp Concordance and Word Profiler Version 4 program. The program generates a list of all words in a given text file, which can be displayed by frequency or in alphabetical order. For each text file, a list of all words by alphabetical order was generated. The frequencies of *all* occurrences of each keyword representing each subtheme were noted down. The concordance program also allows for a listing of all occurrences of a particular word by context, that is, it displays the broader phrase or section of the sentence for every occurrence of the word. From each master file, the context of occurrence of each subtheme keyword was generated. Every individual occurrence of the subtheme keyword was assessed for relevance to ensure whether it conformed to the sense in which the keyword had been defined as representative of a theme. For example, the theme of Rights and

Freedoms included the word "right" as part of the set of keywords collectively representing the theme. In the final sample of text collated from Kashmiri nationalism, the word "right" might, as a hypothetical, be used, for example, in a sentence about the "right of Kashmiris to self-determination." In this case, the occurrence of the word "right" conforms to the sense in which the keyword "right" has been included in the theme of Rights and Freedoms. However, the word "right" might also be used, hypothetically speaking, in the same sample in a distinctly different sense as synonymous with "correct," for example, in a sentence such as "we urge the Government of India to take the right measures." Where the meaning of a word was not apparent from the phrase or part of the sentence displayed by the concordancer, a larger section of contextual text was checked directly from the master file. Appendix E presents the unweighted frequencies of relevant instances of all subtheme keywords and all themes identified from the master files of the three online nationalism samples.

The unweighted frequencies of all themes were converted to weighted frequencies, according to a base of $N = 100$. Since the sample size for each online Indian nationalism is different, weighting each sample to a scale of one hundred allows for a comparison across the same sample size. The weighted figures were rounded off to the nearest whole number. In this manner, final frequencies for all the themes were tabulated for each nationalism. Tables 2 and 3 depict the unweighted and weighted frequencies for all themes for each nationalism. Table 3 indicates the rounded-off frequencies used for the CA in bold and parentheses.

The data in table 3 were entered into the JMP software program. For variable *x*, Online Indian Nationalism, the data type was designated as character, and the modeling type as nominal. Likewise, for variable *y*, Salient Ideological Theme, the data type was designated as character, and the modeling type was ordinal. For each *xy* pair—such as HN-G, HN-RF, and so on; SN-G, SN-RF, and so on; KN-G, KN-RF, and so on—the corresponding frequency was entered in another column, "Count," where the data type was designated as numeric and model type was designated as ordinal.

TABLE 2. Unweighted theme frequencies.

	G	RF	HG	TV	S	U	TK
HN	168	72	3	47	83	9	73
SN	26	126	5	55	0	0	0
KN	74	156	6	50	15	3	3

Note. N for HN = 36, SN = 15, KN = 21.

TABLE 3. Weighted theme summaries.

	G	RF	HG	TV	S	U	TK
HN	466.67 **(467)**	200 **(200)**	8.33 **(8)**	130.55 **(131)**	230.55 **(231)**	25 **(25)**	202.77 **(203)**
SN	173.33 **(173)**	840 **(840)**	33.33 **(33)**	366.66 **(367)**	0 **(0)**	0 **(0)**	0 **(0)**
KN	352.38 **(352)**	742.85 **(743)**	28.57 **(29)**	238.09 **(238)**	71.42 **(71)**	14.28 **(14)**	14.28 **(14)**

Note. N = 100 for HN, SN, and KN.

The "Analyze" function was selected from the toolbar, following which the "Fit Y by X" function was selected. "X, Factor" was specified as "Online Indian Nationalism," "Y, Response" was entered as "Salient Ideological Theme" and the Count was identified as a Frequency variable. The output of the analysis was displayed in two forms, representing a contingency analysis of Salient Ideological Theme by Online Indian Nationalism: first, a mosaic plot that visually depicts the distribution of each theme by each nationalism (figure 1); and second, a contingency table (table 4), which lists the row profiles and column profiles for each category of the variables in terms of percentages. Following this, the correspondence analysis function was also selected, and the output was displayed in the form of a two-dimensional diagram (figure 2).

FIGURE 1. Mosaic plot for contingency analysis of Online Indian Nationalism by Salient Ideological Theme.

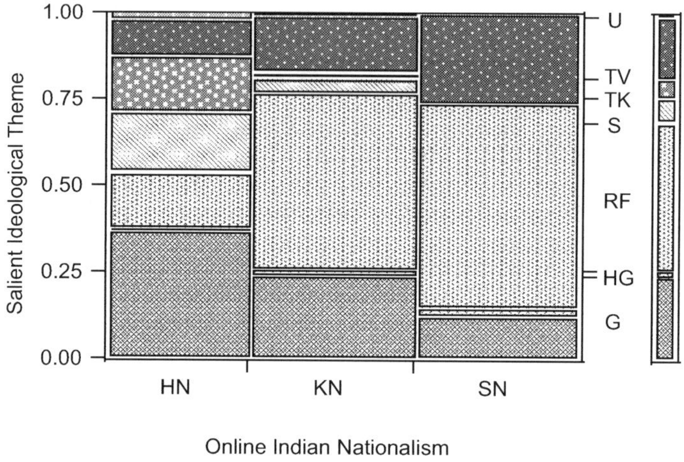

Freq: Count

TABLE 4. Contingency table for Online Indian Nationalism by Salient Ideological Theme.

Count Total % Col % Row %	**G**	**HG**	**RF**	**S**	**TK**	**TV**	**U**	
HN	467	8	200	231	203	131	25	1265
	11.28	0.19	4.83	5.58	4.90	3.17	0.60	30.56
	47.08	11.43	11.22	76.49	93.55	17.80	64.10	
	36.92	0.63	15.81	18.26	16.05	10.36	1.98	
KN	352	29	743	71	14	238	14	1461
	8.50	0.70	17.95	1.72	0.34	5.75	0.34	35.30
	35.48	41.43	41.67	23.51	6.45	32.34	35.90	
	24.09	1.98	50.86	4.86	0.96	16.29	0.96	
SN	173	33	840	0	0	367	0	1413
	4.18	0.80	20.29	0.00	0.00	8.87	0.00	34.14
	17.44	47.14	47.11	0.00	0.00	49.86	0.00	
	12.24	2.34	59.45	0.00	0.00	25.97	0.00	
	992	70	1783	302	217	736	39	4139
	23.97	1.69	43.08	7.30	5.24	17.78	0.94	

FIGURE 2. Correspondence analysis.

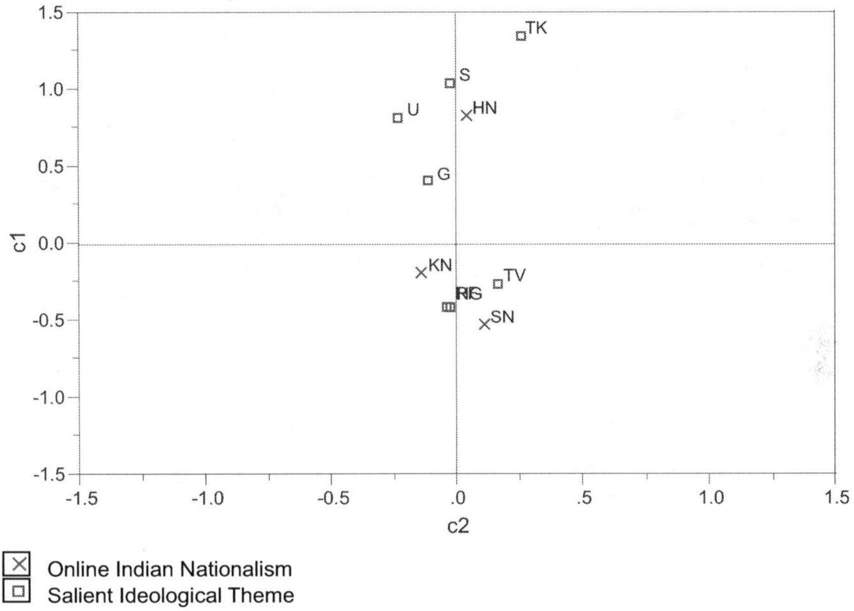

TABLE 5. Tests.

Source	DF	-LogLike	RSquare (U)
Model	12	740.6872	0.1217
Error	4121	5346.8491	
C. Total	4133	6087.5363	
N	4139		

Test	ChiSquare	Prob > ChiSq
Likelihood Ratio	1481.374	< .0001
Pearson	1361.278	< .0001

TABLE 6. Details of correspondence analysis.

Singular Value	Inertia	Portion	Cumulative
0.56341	0.31743	0.9652	0.9652
0.10705	0.01146	0.0348	1.0000
Online Indian Nationalism		**c1**	**c2**
HN		0.8231	0.0398
KN		–0.1985	–0.1399
SN		–0.5316	0.1091
Salient Ideological Theme		**c1**	**c2**
G		0.398	–0.1113
HG		–0.424	–0.0187
RF		–0.427	–0.0229
S		1.035	–0.0233
TK		1.344	0.2630
TV		–0.324	0.1516
U		0.810	–0.2312

INTERPRETATION OF RESULTS

The row profiles and column profiles of each variable category are presented in the preceding contingency table (table 4). The contingency table details the proportionate percentage of each variable category frequency to the total count of the table, as well as the proportionate percentage of each frequency to the row and column totals. An example will clarify this. The frequency count for Globality for Hindu Nationalism (HN-G) is 467. This amounts to 11.28 percent of the total table sum of 4,139. The frequency count of 467 amounts to 47.08 percent of the column total of 1,265, and it amounts to 36.92 percent of the row total of 992. In other words, the incidence of Globality in online Hindu nationalism accounts for 47.08 percent of all incidences of Globality across the three online nationalisms and for 36.92 percent of incidences of all themes within online Hindu nationalism. The results of the correspondence analysis are displayed in visual form in figure 2, which allows us to ascertain whether the hypothesis is confirmed, provide a map of the spread of different themes with regard to the nationalisms, and interpret the results. The three online Indian nationalisms are represented as ⊠, while the seven salient ideological themes are represented as ⊡. The themes of Rights and Freedoms (RF) and Holocaust and Genocide (HG) are placed very close to each other and practically overlap, so the letters designating each theme appear somewhat unclear.

The results indicate that the theme of Scientificity (S) is most strongly associated with online Hindu nationalism and is weakly associated (spatially significantly removed) from both online Sikh nationalism and online Kashmiri nationalism. As indicated by the row profile in table 4, the frequency of the theme of Scientificity accounts for 18.26 percent of the occurrence of all themes for online Hindu nationalism, second only to the thematic emphasis on Globality (36.92 percent) in online Hindu nationalism. The column profile in table 4 indicates that incidence of scientificity in online Hindu nationalism accounts for 76.49 percent of all incidences of the theme across the three nationalisms. Truth and Knowledge (TK) and Universality (U) are also most strongly associated with online Hindu nationalism and far removed from online Sikh nationalism and online Kashmiri nationalism. The hypothesis is accordingly confirmed with regard to the relationship between Scientificity and online Hindu nationalism, as indeed is the proposition that an emphasis on the themes of Universality and Truth and Knowledge is peculiar to online Hindu nationalism as key aspects of the ideological discourse of the nationalism.

The theme of Globality (G) is placed midway between online Kashmiri nationalism and Hindu nationalism and somewhat further from online Sikh nationalism. This may be explained by the fact that the theme of Globality is more relevant for the first two online nationalisms—though for different reasons—than for online Sikh nationalism. As explained earlier, the diaspora of Hindus is a particularly important constituency for Hindu nationalism. The medium of the Internet is considered a crucial arena and instrument through which a transnational population of Hindu nationalist sympathizers present their claims about the Hindu nation. The global nature of the population of Hindu nationalist proponents in cyberspace may itself be partly reflected as an emphasis on globality in Hindu nationalist discourse online. More importantly, a central claim of Hindu nationalist ideology in its current avatar in post-liberalization India is that Hinduism is perfectly compatible with the demands of global capitalism and, in fact, is ideally suited as an overarching political-cultural framework for an India rapidly being integrated into a globalized world. These factors explain the emphasis on globality for online Hindu nationalism.

For online Kashmiri nationalism, the emphasis on globality is rooted in different motivations. The claim of self-determination is central to the movement for Kashmiri independence. This claim is not accepted by the Indian state in its negotiations with representatives of the Kashmiri secessionist movement. The representatives of the segment of the Kashmiri population that seeks independence from both India and Pakistan would have adopted a strategy of internationalizing the issue. Such adherents and sympathizers of the movement would seek to mobilize support from an international, global community by framing the case for Kashmiri independence in terms of the fundamental and universal right of all peoples of the globe to self-determination. Also, from its very inception, historically, the Kashmir conflict has been defined by Pakistan as an international issue, even as India has repeatedly insisted that it is a bilateral issue. Although the Simla Agreement signed between India and Pakistan in 1972 proposed to solve the dispute through bilateral negotiations, the UN Charter was invoked as the broader framework for relations between the countries.[66]

The difference between Sikh nationalism and Kashmiri nationalism may be partly explained by the fact that the movement for an independent Khalistan has largely been quashed within India, although Sikh organizations overseas, such as the Dal Khalsa, Babbar Khalsa, Council of Khalistan, and the Sikh Federation, continue to promote the cause.[67] Arguably, such groups have not managed to mobilize support from the international community, since several have been banned or designated as terrorist organizations by the United States and Britain.[68] While several militant Kashmiri organizations or militant organizations operating in Kashmir have also been designated as terrorist groups by the United States and United Kingdom, the Kashmir issue is still generally treated as a legitimate international political conflict. In contrast, the Khalistan issue appears to be treated as a less legitimate and viable cause by the global or international community in general. The motivations for online Sikh nationalism to emphasize the quality of globality appear weaker than corresponding motivations in the case of Kashmir. Similarly, the global response to the conflicts is less enthusiastic in the case of the Khalistan cause as opposed to the cause for an independent Kashmir. However, there is still some emphasis on the theme of Globality in online Sikh nationalism, as the column profile figure of 17.44 percent and the row profile figure of 12.24 percent indicate. It is the third most dominant theme for online Sikh nationalism out of the seven themes. The fact that the theme of Globality is prominent to varying extents in all nationalisms appears to confirm the general assumption based on which the theme was initially selected: the idea that nationalisms would articulate themselves in cyberspace as *global* issues of relevance to the international community at large as well for diasporic ethnic, cultural, or religious groups dispersed across the world to speak to the potentially global audience using the Internet.

The theme of Rights and Freedoms (RF) is more or less equally closely placed between online Sikh nationalism and online Kashmiri nationalism and is distantly located from online Hindu nationalism. The location of the theme reflects a relatively strong association with the first two online nationalisms as compared with online Hindu nationalism. This can be explained by the fact that a discourse of rights and freedoms is far more central to the secessionist movements seeking independence from the Indian state. The rights and freedoms of the members of the minority religious or ethnic communities of Sikhs and Kashmiris (whether as individuals or groups), as articulated in the Indian constitution or international treaties, provide a strong basis for criticism of the Indian state and for establishing the legitimacy of secessionist nationalist causes. In contrast, the Hindu nationalist movement does not seek to criticize the Indian state for violating the rights and freedoms of Hindus. The Hindu nationalist movement does not seek to overthrow the Indian state, nor is locked with it in permanent conflict. Rather, the Hindu nationalist movement seeks to control the Indian state through the structures of representative democracy. While a key claim in Hindu nationalist discourse is that the Indian state offers preferential treatment to religious minorities in comparison to its treatment of the majority Hindu population, the Hindu nationalist movement seeks to correct this perceived imbalance through political and legislative change. It is important to note here that the Hindu nationalist BJP has never gained a significant enough majority in any general election to form the Indian government by itself. While it has spearheaded Indian governments in the 1990s, it has always done so as part of a coalition with other parties, often on the basis of a compromise that it will not enforce particular policy objectives on its agenda. Rather than challenging the constitution, democracy, or the Indian state, Hindu nationalist organizations seek to mobilize Hindus to vote for the BJP in large enough numbers such that a BJP government might come to power with an absolute majority in the Indian parliament. The long-term strategic objective behind such mobilization is that a BJP government will be able to pass changes in law and amend the Indian constitution toward ensuring that "Hindu interests" are protected. Also, given that Hindus are an overwhelming majority in India, accounting for over 80 percent of the population, Hindu nationalist discourse cannot make the claim that Sikhs or Kashmiri Muslims are oppressing the Indian population at large, although Hindu nationalists do point to the persecution of Hindus in Kashmir. Neither can Hindu nationalist discourse claim that the Indian state is controlled by representatives of religious minorities. Sikhs or Kashmiri Muslims, however, can make the argument that the Indian government is largely Hindu controlled and is indifferent to, and oppressive of, religious minorities in general.

The theme of Terror and Violence (TV) is located most closely to online Sikh nationalism, fairly close to online Kashmiri nationalism, and distant from online Hindu nationalism. The associations can be explained by the fact that both secessionist movements for Kashmir and Khalistan have been engaged in militant conflict with the Indian state and, in each case, the Indian state has used significant military force to combat the secessionists. This conflict has been marked by human rights violations and brutalities on the part of the Indian army and police forces in both zones of conflict, as recorded by human rights organizations.[69] The Hindu nationalist movement, in contrast, is not engaged in military conflict with the Indian state. There have been instances when supporters of Hindu nationalist causes have clearly violated the law and fomented violence, such as in the case of the Babri Masjid demolition in 1992 and during communal riots in Bombay in 1992–1993 or in Gujarat in 2002. Yet, in such cases, far from violating the rights of Hindus, the Indian state has been lax in holding the Hindu nationalist organizations responsible for the violence accountable to justice. On the contrary, the state and its instruments, such as the police, have been guilty of complicity in the violence against religious minorities. The state has been equally guilty of deliberate inaction to prosecute guilty members of the majority Hindu community, especially when such individuals are affiliated with Hindu nationalist organizations. Regardless of whether a Hindu nationalist party has been part of the government or not, the actions of the Indian state in preventing terror and violence against minorities in the last two decades stand as a shameful failure.

It is true that Hindu nationalist groups have drawn attention to violence perpetrated against Hindus by Muslims in Kashmir or Islamic terrorism in India. Similar claims are also made in Hindu nationalist discourse about the violent nature of Islam in general. The contingency table (table 4) does indicate some prevalence of the theme of Terror and Violence in online Hindu nationalism, which may be explained by these factors. The incidence of the theme of Terror and Violence in online Hindu nationalism amounts to a proportion of 17.80 percent of the total occurrence of the theme across the three nationalisms. It amounts to a proportion of 10.36 per cent out of the occurrence of all seven themes in online Hindu nationalism. But given the fact that over 80 percent of India's population is Hindu, it is unlikely that Hindu nationalists would strongly and persistently argue that the 13.4 percent Muslim population or 1.9 percent Sikh population is permanently tyrannizing or oppressing the Hindu majority in India.[70] To sum up, the preceding figures reveal that the theme of Terror and Violence, while also important to online Hindu nationalism, is not as central to it as it to the other discourses.

There is some difference in the degree of association of the theme of Terror and Violence with online Sikh nationalism and association of the theme with online Kashmiri nationalism, although the theme is clearly relevant for both nationalisms. One can speculate about the reason for this difference. As previously suggested, the cause of Kashmiri independence appears to be a stronger political and social force in India than the cause of Khalistan. The Kashmiri cause is also more visible in the domain of international politics. It may be the case that the Khalistan movement emphasizes the theme of Terror and Violence to a relatively greater degree in presenting its case, since it cannot emphasize other themes such as Globality to the same extent.

Finally, the theme of Holocaust and Genocide (HG) is also placed equally between online Sikh nationalism and online Kashmiri nationalism and distant from online Hindu nationalism. But we may note from the contingency table that the theme of Holocaust and Genocide accounts for a very minimal weight in terms of the row and column profiles. While in relative terms it is more strongly associated with online Sikh and Kashmiri nationalisms as compared to online Hindu nationalism, the prevalence of the theme in online nationalist discourse at large does not appear to be significant.

In conclusion, the CA confirms the hypothesis that scientificity is most strongly associated with online Hindu nationalism as compared to other online Indian nationalisms. Scientificity is also the second most dominant theme in online Hindu nationalism, after the theme of globality. Other key substantive elements of Hindu nationalist discourse, such as the emphases on truth and knowledge and universality, are also shown to be most strongly associated with online Hindu nationalism. Other themes vary in their associations with the three nationalisms as previously explained. We can conclude, therefore, that the emphasis on scientificity in online Hindu nationalism is not an artifact or outcome of the online virtual status of the discourse but, rather, is reflective of the ideological centrality of the theme in the discourse. The same argument may be made with regard to the association of certain themes for online Sikh nationalism and online Kashmiri nationalism.

But it is not the case that the Internet—or the online—status of the three nationalisms analyzed here does not have any impact on the forms taken by them in cyberspace. The emphasis on globality, present in all, for example, can be interpreted as a function of the online status of the discourses. Globality is the most dominant theme for online Hindu nationalism out of all seven themes, with the row profile figure at 36.92 percent. It is the second most dominant theme for online Kashmiri nationalism, with the row profile figure of 24.09 percent, and the third most dominant theme for online Sikh nationalism, with a row profile figure of 12.24 percent. It is unlikely that Hindu nationalism, Sikh nationalism, or Kashmiri nationalism will emphasize globality when they speak to populations in India whose concerns may be predominantly local or whose socio-economic profile may differ significantly from the online constituencies of each discourse. Similarly, it is unlikely that the Kashmiri and Sikh nationalist movements will refer to the UN Charter on human rights or international covenants when addressing local populations about the violence committed by the Indian state.

One can speculate, then, that as these nationalisms express themselves online, they retain their key ideological concerns but also add and foreground other concerns that are appropriate to the wider economic, social, and political context of Internet use. Equally, the key ideological concerns may be reformulated in an idiom of global relevance, as suggested earlier. For example, it is likely that Hindu nationalism emphasizes scientificity in a *particular* mode when it is online, such that the scientificity of Hinduism itself is described as a necessary quality for survival in the world of global capitalist modernity represented by technologies such as the Internet. The final chapter addresses these and related questions through a reflection on the reformulation of identities in the globalized, technologically-driven world of the twentieth century and the shadows of that world where alternate imaginings of self and society also glimmer.

ENDNOTES

1. This chapter represents a shift in the mode of analysis for evaluating the relationship between technology and nationalism and for assessing the characteristics of technocultural Hindu nationalism or Hindu nationalism online.
2. Sten-Erik Clausen, *Applied Correspondence Analysis: An Introduction* (Thousand Oaks, CA: Sage, 1998), 1.
3. *Google*, http://www.google.com (accessed May 1, 2005).
4. The statistical software package used is JMP Professional, Version 5.0.1.2. See *JMP*, http://www.jmp.com (accessed May 1, 2005).
5. Nitish Singh and Daniel W. Baack, "Web Site Adaptation: A Cross-Cultural Comparison of U. S. and Mexican Web Sites," *Journal of Computer-Mediated Communication* 9, no. 4 (July 2004), available from http://jcmc.indiana.edu/vol9/issue4/singh_baack.html (accessed July 11, 2005).
6. Ibid. The typology of cultural values is drawn from Geert Hofstede, *Culture's Consequences: International Differences in Work-Related Values* (Thousand Oaks, CA: Sage, 1980). The authors selected Hofstede's typology after a detailed examination of various cultural typologies used in business literature.
7. Singh and Baack, "Web Site Adaptation."
8. Ibid.
9. Ibid.
10. Luca Tateo, "The Italian Extreme Right On-Line Network: An Exploratory Study Using an Integrated Social Network Analysis and Content Analysis Approach," *Journal of Computer-Mediated Communication* 10, no. 2 (January 2005), http://jcmc.indiana.edu/vol10/issue2/tateo.html (accessed July 10, 2005).
11. Ibid.
12. Ibid.
13. Nicola Doring, "Personal Home Pages on the Web: A Review of Research," *Journal of Computer-Mediated Communication* 7, no. 3 (April 2002), http://jcmc.indiana.edu/vol7/issue3/doering.html (accessed July 7, 2005).
14. Ibid.
15. Ibid.
16. Richard Holt, *Dialogue on the Internet: Language, Civic Identity and Computer-Mediated Communication* (Westport, CT: Praeger, 2004).

17. Ibid., 131.
18. Ibid., 131.
19. Daniel Marschall, "Ideological Discourses in the Making of Internet Career Sites," *Journal of Computer-Mediated Communication* 7, no. 4 (July 2002), http://jcmc.indiana.edu/vol7/issue4/marschall.html (accessed July 10, 2005).
20. Ibid.
21. Ibid.
22. Holt, *Dialogue on the Internet*, 129.
23. Ibid., 133.
24. Ibid., 133.
25. Harnik Deol, *Religion and Nationalism in India: The Case of the Punjab* (New York: Routledge, 2000), 11.
26. Ibid., 1.
27. As per the 2001 census, the Hindu population of India was approximately 827.6 million, which represented 80.5 percent of the total Indian population. Government of India. Office of the Registrar General & Census Commissioner, India, "Census and You—Religion" (accessed June 22, 2008).
28. See Deol, *Religion and Nationalism in India*, esp. chap. 4, "The Rise of Sikh National Consciousness (1947–95)," 92–124, for an excellent treatment of Sikh nationalism.
29. Deol, *Religion and Nationalism in India*, 4.
30. Harjot Oberoi, "From Punjab to 'Khalistan': Territoriality and Metacommentary," *Pacific Affairs* 60, no. 1 (Spring 1987): 27.
31. Ibid., 30.
32. Deol, *Religion and Nationalism in India*, 93.
33. Ibid., 94.
34. Ibid., 94, 96–97.
35. Ibid., 104.
36. Khilnani, *The Idea of India*, 51.
37. Ibid., 51–52.
38. Ibid., 52.
39. Oberoi, "From Punjab to 'Khalistan,'" 39.
40. Rajesh Deol, "Cry for Khalistan Again," *Deccan Herald*, January 18, 2004, http://www.deccanherald.com/deccanherald/jan182004/fp5.asp (accessed May 4, 2005).
41. For a good treatment of the Kashmir conflict and secessionist movement, see Sumit Ganguly, "Explaining the Kashmir Insurgency: Political Moblization and Institutional Decay," *International Security* 21, no. 2 (Autumn 1996): 76–107; and Reeta Chowdhury Tremblay, "Nation, Identity and the

Intervening Role of the State: A Study of the Secessionist Movement in Kashmir," *Pacific Affairs* 69, no. 4 (Winter 1996–1997): 471–497.

42. T. N. Madan, "Religions of India: Plurality and Pluralism," in *Religious Pluralism in South Asia and Europe*, ed. Jamal Malik and Helmut Reifeld (New Delhi: Oxford University Press, 2005), 45. The union territory of Lakshwadeep is the only other Muslim majority region in India (ibid.).
43. See "Who Are the Kashmir Militants?" *BBC News*, February 19, 2003, http://news.bbc.co.uk/1/hi/world/south_asia/1719612.stm#map (accessed July 17, 2005).
44. See "National Self Determination," *Khalistan, the New Global Reality*, http://www.khalistan.net/washtimes.htm (accessed June 10, 2005); and "Sikh Struggle for Independence," *Khalistan, the New Global Reality*, http://www.khalistan.net/unpo.htm (accessed June 10, 2005.
45. See the *Hindu Holocaust Museum* (accessed July 2, 2005); and the *Online Hindu Holocaust Museum* (accessed July 2, 2005).
46. See "Islamic Terrorism and Genocide of Kashmiri Pandits," *Virtual Homeland of Kashmiri Pandits*, http://www.kashmiri-pandit.org/sundry/genocide.html (accessed July 3, 2005).
47. See, for example, Koenraad Elst, *The Koenraad Elst Site*, http://koenraadelst.voiceofdharma.com/ (accessed June 17, 2005).
48. *Islam/Muslims in Kashmir*, http://www.geocities.com/WestHollywood/Park/6443/Kashmir/ (accessed June 10, 2005).
49. As of March 17, 2006, the Web site has been renamed as *Hindutva Brotherhood: Kashmiri Hindu Liberation Army* and the URL http://www.geocities.com/jairama1/redirects to the URL http://www.hindurashtra.org/. See *Hindutva Brotherhood: Kashmiri Hindu Liberation Army*, http://www.geocities.com/jairama1/and http://www.hindurashtra.org/ (accessed March 17, 2006).
50. The Concapp Concordance and Word Profiler Version 4 software program is available free on the Web at the Web site of the Edict Virtual Language Center, an electronic language resource. See "ConcApp Concordancing Programs," *Edict Virtual Language Center*, http://www.edict.com.hk/pub/concapp/ (accessed April 3, 2005).
51. This is true, of course, as of the writing of this chapter and would have changed subsequently. See the following for details on Google's search technology: "Our Search: Google Technology—Benefits of Google?" *Google*, http://www.google.com/technology/whyuse.html (accessed April 4, 2005); "Google Information for Webmasters," *Google*, http://www.google.com/webmasters/4.html (accessed April 4, 2005); "Google Corporate Information,

Technology Overview," *Google*, http://www.google.com/corporate/tech.html (accessed April 4, 2005); and "Our Search: Google Technology—Why Use Google?" *Google*, http://www.google.com/technology/index.html (accessed April 4, 2005).

It should also be acknowledged that scholars have raised some important questions about Google's ranking system, such as possible factors that may compromise its efficiency and the phenomenon of "googlearchy," that is, the extent to which a few powerful sites monopolize the flow of information on the World Wide Web. See Bing Pan et al., "In Google We Trust: Users' Decisions on Rank, Position, and Relevance," *Journal of Computer-Mediated Communication* 12, no. 3, article 3 (2007), http://jcmc.indiana.edu/vol12/issue3/pan.html (accessed October 12, 2007); Matthew Hindman, Kostas Tsioutsiouliklis, and Judy A. Johnson, *"Googlearchy": How a Few Heavily-Linked Sites Dominate the Web* (2003), available from http://www.cs.princeton.edu/~kt/mpsa03.pdf (accessed October 12, 2007).

52. *World Sikh Council*, http://www.worldsikhcouncil.org/ (accessed April 7, 2005). See the press release, World Sikh Council, "WSC-AR Condemns Denial of Justice to Sikh Victims of November 1984 in India; Appeals to World Community and UN," *World Sikh Council* (October 28, 2004), http://www.worldsikhcouncil.org/press/2004/2004_Nov_PressRelease.pdf (accessed April 7, 2005).
53. See the section, "Links," *World Sikh Council*, http://www.worldsikhcouncil.org/docs/links_sites.html (accessed April 7, 2005).
54. "About This Site," *Hindu Universe* (accessed February 19, 2005).
55. Clausen, *Applied Correspondence Analysis*, 25.
56. Michael Greenacre, "Correspondence Analysis and its Interpretation," in *Correspondence Analysis in the Social Sciences: Recent Developments and Applications*, ed. Michael Greenacre and Jorg Blasius (New York: Harcourt Brace, 1994), 1.
57. Clausen, *Applied Correspondence Analysis*, 5.
58. Bourdieu, *Distinction*.
59. Greenacre, "Correspondence Analysis and its Interpretation," 8.
60. Ibid., 8–9.
61. Ibid., 10.
62. Michael Greenacre, *Correspondence Analysis in Practice* (San Diego: Academic Press, 1993), 19.
63. Ibid., 25–26.
64. Ibid., 31.
65. Clausen, *Applied Correspondence Analysis*, 16.

66. Embassy of India, Washington, DC, "Simla Agreement," *Embassy of India*, http://www.indianembassy.org/policy/Kashmir/shimla.htm (accessed June 10, 2005).
67. Rajesh Deol, "Cry for Khalistan Again" (accessed May 4, 2005).
68. Ibid.
69. See, for instance, the following reports: Human Rights Watch, *Behind the Kashmir Conflict: Abuses by Indian Security Forces and Militant Groups Continue* (July 1999), http://www.hrw.org/reports/1999/kashmir/abuses.htm (accessed June 1, 2005); Human Rights Watch, *India's Secret Army in Kashmir: New Patterns of Abuse Emerge in the Conflict* 8, no. 4 (C) (May 1996), http://www.hrw.org/campaigns/kashmir/1996/ (accessed June 10, 2005); and Human Rights Watch, *India: Arms and Abuses in Indian Punjab and Kashmir* 6, no. 10, (September 1994), http://www.hrw.org/campaigns/kashmir/1994/ (accessed June 11, 2005).
70. Government of India, Office of the Registrar General, "Census and You—Religion" (accessed June 22, 2008).

Chapter 7

Conclusion: Identities in the Era of Globalization and the Internet

In this concluding chapter, I theorize the Internet as an embodiment of social space and reflect on the possibilities offered by the Internet for reimagining Indian identities in light of the historical association between nationalism and technology. I address the complexities of the production, representation, and consumption of online discourse in general, including but not limited to online Hindu nationalism.[1] The first section of the chapter offers a close reading of a statement about Indian identity presented in an Indian government document on the World Wide Web. The second section analyzes some academic perspectives on the social impact of the Internet. The third section employs Bourdieu's concepts in theorizing the dynamics of the online representation and reconfiguration of social space. The final section briefly compares online Hindu nationalism and alternate visions of Indian identities in cyberspace.[2]

Here I also comment on some strengths and limitations of Bourdieu's theoretical framework.

READING "OFFICIAL" INDIAN IDENTITY ON THE INTERNET

At some point of time in the last few years, the NRI and PIO division of the Ministry of External Affairs of the Government of India set up a Web site dedicated to the Indian diaspora.[3] The Web site includes a report of a high level committee that was specially formed to study issues concerning, and of concern to, the twenty million or so NRIs and PIOs dispersed all over the world.[4] The construction of the Indianness of the diaspora, defined in various ways in the foreword of the report, is deserving of attention.[5] The Indian diaspora is described in very broad terms as constituted of "Indians who migrated to different parts of the world and *generally maintained* their identity."[6] At the same time, the chief representatives of the diaspora appear largely to be successful professionals; these Indian achievers cover the categories of "entrepreneurs, workers, traders, teachers, researchers, inventors, lawyers, engineers, managers and administrators."[7] And, additionally, the diaspora is defined through a mutually symbiotic relationship with the Indian state: it both reflects and is reflected in the Indian nation. It is emphasized that "the destiny of the Indian Diaspora has always been intertwined with the fortunes of India. India's progress and prosperity is located in the stature accorded to her Diaspora wherever it is located."[8] While the relationship between nation and diaspora earlier took the form of a "hub and spoke relationship," today this relationship is being recast along the lines of a "network design of a 'web relationship.'"[9] As is not too difficult to discern, the current relationship between the Indian nation and diaspora is described in terms of a metaphor that evokes the Internet and the World Wide Web.

Whether intended or not, the technological metaphor is instructive in revealing where exactly the importance of the Indian diaspora lies for the Indian nation-state and vice versa. Sentimental invocations of the link between the mother country and her globally dispersed progeny notwithstanding, the compact between the Indian nation and diaspora is founded on the technological and economic accomplishments of the Indian diaspora and, relatedly, the rise of India as a key player in the global technological industry and the world economy. As the report stresses, the Indian diaspora has "transformed the economies" of the various countries in which members of the diaspora find themselves, and in several of these countries, the diaspora "has surpassed the per capita income of the indigenous population."[10] The diaspora has also played a "leading role in the global technological revolution."[11] India's visibility in the global field of information technology and the nation's rise to relatively greater prominence in the arena of international relations have bathed the diaspora in a new light of respectability. In turn, the diaspora's expertise in technology can help India prosper as a technological and economic powerhouse. Through its increased political visibility in various countries, the diaspora is well placed to foster beneficial relations between India and these nations.[12]

Earlier chapters have outlined the broader historical reasons, as well as the more immediate, post-1991 context, that explain the importance of a diasporic population of technocratic professionals for the Indian state. These arguments do not need to be revisited here, but some relevant points may be noted. In the report, Indian national and cultural identity—which binds the Indian nation-state with the diaspora—is reimagined through the prism of economic-technological markers. Technological expertise, generation of capital, and professional capability are redefined as indicators of cultural and national values, reflected at once in the personal achievements of diasporic Indians and in the policies of the Indian state. Concurrently, the rephrasing of Indian identity in official discourse is significantly informed by structural changes at the national and global level, such as technological development and, more generally, India's integration into a globalized economy and world. And, interestingly, the report presents its propositions about national and diasporic Indian identity as credible historical claims, much in the same manner that Hindu nationalist Web sites present their arguments about Indian identity as historical arguments.[13] In both cases, there is a reductive rearticulation of Indian identity, which reflects the interests of the particular constituencies deemed important by the respective producers of the two types of Web sites, the Government of India and Hindu nationalists. Just as the claims of Hindu nationalists are not borne out by the realities of Indian life, the picture of state-diaspora relations presented on the *Indian Diaspora* Web site is not necessarily accurate. For example, the claim that the interests of the Indian diaspora have always been linked with those of the Indian state is contradicted by the reaction of expatriate Indians to India's economic crisis in 1991. The withdrawal of investments from the Indian economy by NRIs was one cause of the balance-of-payments crisis faced by the Indian state in 1991.[14] Hindu nationalist Web sites emphasize a culturally exclusivist idea of Indian identity as rooted in Hindu values, to demonstrate how the Hindu-Indian ethos is perfectly suited to exploit the opportunities available in an economically and technologically networked globalized world. Similarly, the *Indian Diaspora* Web site codes Indian cultural values as fundamental qualities possessed by Indians individually and collectively that are ideal for meeting the demands of the present global marketplace. An affluent resident and diasporic Indian community with sympathies to Hindu nationalist ideology—its ranks swelled with technology professionals—is the prime audience of Hindu nationalist Web sites. An affluent NRI community appears to be the prime audience of the diaspora Web site.

On both Hindu nationalist Web sites and the *Indian Diaspora* Web site, the representation of Indian identity reflects the impact of wider national and global developments in politics, economy, and technology, and, equally, mirrors the importance of particular audiences and constituencies. At the same time, the cyber-representation of Indianness can itself be seen as a particular economic, technological, and cultural intervention in social space. The fact that the target constituencies of both the Hindu nationalist Web sites and the *Indian Diaspora* Web site are significantly numbered by Internet users no doubt motivates the respective producers of the online resources to affirm as valuable the professional and ideological choices of the respective audiences. Castells' theorization that locates Internet practices in the context of social practice at large is useful here for grasping this twofold schematic of the Internet as simultaneous representation and intervention into social space.[15] On the one hand, as Castells points out, the Internet is continuous with society itself: "[i]t is an extension of life as it is, in all its dimensions and all its modalities" and "has been appropriated by social practice, in all its diversity."[16] On the other hand, "this appropriation does have specific effects on social practice itself."[17] This points to the broad question addressed in this chapter: how does the Internet—simultaneously an instrument of technology, a form of media, and a social-cultural space—both reflect and reconfigure the processes through which social identities are produced, contested, and reshaped? A related area of inquiry is how the reconfiguration of social space by the Internet operates with regard to the online expression of Indian identities, as seen in online Hindu nationalism and the *Indian Diaspora* Web site. In engaging with these questions, the next section reviews some perspectives on the social impact of the Internet.

Perspectives on the Internet and Social Space

From cyberspace to virtual community, cybertopia to virtual nation, and home pages to World Wide Web, descriptions of Internet usage conceptualize the experience of being online as akin to inhabiting a residence, locality, or larger geographical territory. Such territorially demarcated locations in cyberspace are simultaneously social and cultural spaces, framed by commonalities of identity such as ethnicity or language, shared interests such as sports or news, or functional actions such as paying a bill. The vocabulary of Internet usage normatively defines social practice as a topographical act: to be online is to navigate a microcosm of the social world; it is to occupy the virtual equivalent of different types of spaces in the offline world, personal or professional, familial or national, or commercial or leisurely. At the same time, given its increasing reach and importance across diverse areas of social existence, the Internet is seen as altering the very map of social space. For instance, the oft-mentioned digital divide between those who have access to the Internet and those who do not is viewed as further deepening the significant gulf that exists between the two sides in terms of resources and opportunities. Given the profusion of tropes of location, place, and territory in discourse about the Internet, the general notion of "space" is helpful as a basic conceptual frame for assessing perspectives on the social impact and significance of the Internet.

It is instructive to begin by reviewing two diametrically opposed positions on the significance of the Internet, the utopian and dystopian, which represent the two ends of a spectrum of perspectives on the Internet. The utopian view is confident about the radical emancipatory and progressive possibilities offered by the Internet, while the dystopian view sees the Internet as reflecting existing social inequalities as well as contributing to the exacerbation of such inequities. Ebo uses the spatial metaphors of the "cybertopia" and the "cyberghetto" to describe these two perspectives.[18] The cybertopia view sees the Internet as a "great equalizer because virtual communities do not use disenfranchising criteria for community building."[19] In the cyberghetto reading, "virtual communities are just another dimension of traditional communal relationships with the same built-in biases of race, class, and gender."[20] By implication, the Internet reinforces such biases and may contribute to their further entrenchment in society.

According to the utopian perspective, the Internet offers radically transformed experiences of social space, which, in turn, can refashion the structures of social space in the "offline" world. The Internet enables the emergence of new forms of identity and community as well as engenders hitherto unknown possibilities for a radically democratic politics. As the central agency and symbol of a virtual social revolution, the Internet materializes utopian possibilities that exist in the current world—although these possibilities may be latent, submerged, and dispersed—and makes them accessible to a global community in an egalitarian and universally accessible mode. Rheingold's text, *The Virtual Community: Homesteading on the Electronic Frontier*, is a classic reading exemplifying the utopian view.[21] A celebratory work on the Internet first published in 1993, the text seeks to explain how virtual communities "are likely to change our experience of the real world, as individuals and communities."[22] Rheingold asserts that interaction in electronic forums like the WELL (Whole Earth 'Lectronic Link) is characterized by free exchange of ideas and discussion as well as values of support, kinship, and sympathy. Electronic networks also possess enormous potential for positive political change. They are inherently democratic in structure in that they vest political power primarily in the people. Virtual communities promise a "third place," aside from work and home, where the frayed fabric of sociality, a casualty of the brutal functionalism of the modern world, can be rewoven into wholeness.[23] In a similar vein, Poster argues that the Internet offers the possibility of a postmodern politics, since online communication creates the conditions for the consolidation and emergence of an antifoundational democratic intersubjectivity.[24] This subjectivity is constituted "because the acts of discourse are not limited to one-way address and not constrained by the gender and ethnic traces inscribed in face to face communication."[25] Online, the people are the producers of opinions and the controllers of the "apparatuses of cultural productions."[26] As a result, political authority functions online in an altogether different manner from the real world. On the decentralized and nonhierarchized Internet, authority loses its charismatic power, the "aura," which can no longer legitimize it.[27] Poster concedes that in cyberspace, "asymmetries emerge which could be termed 'political inequalities,'" yet compared to the material world, there is a radical dismantling of hierarchies of race, class, gender, age, and status online.[28] Located in a transcendent space that is relatively free from the constraints of social power relations, Internet communities function as sites of resistance to the dominant modes of sociality in modern society. Virtual communities can accordingly be interpreted as a "domain of uncoerced conversation oriented towards a pragmatic accord," a kind of Habermasian public sphere.[29]

Both texts are early and important reflections on the possibilities of the Internet and offer valuable insights into the communicative potentialities offered by the Internet. Their predictions, however, may need to be qualified. Rheingold's metaphor of the virtual community can be critiqued as a nostalgic idealization of community. In the revised edition of the book, Rheingold does offer a "rethinking" of virtual communities, acknowledging that this discussion is meaningful only as part of a larger debate about developments in communication that should be located in historical context.[30] Rheingold also acknowledges the fact of unequal access to the Internet. But one crucial issue that arises here is whether access to the Internet might itself be a principle of *dividing* communities both within and across societies? One could argue that virtual communities, which may well embody positive values, are created through the forcible creation of other "communities" of those excluded from participation. This factor must be considered in any assessment of the mode in which the virtual community represents a move toward radical political equality or democracy.

Likewise, Poster's conceptualization of the virtual public sphere begs the question of who can participate in this virtual social space; the objections here are the same that can be made with regard to notions of the public sphere in general. Participation in the virtual public sphere presumes, on behalf of participants, ease of access to the Internet, technological literacy, and the ability to write. The presumption of the ability to write reveals, in turn, the privileging of writing over speech, and the privileging of the ability to write in English. The argument that differences in gender, class, and race are erased on the Internet takes for granted that the main mode of communication on the Internet will be writing. This necessarily excludes those who cannot write. However, if the same status were to be granted to speech and images as writing, then the very basis of the virtual public sphere might collapse. For the differences of race, class, gender, and status that get erased in written Internet communication might be relatively easier to discern. Also, well after Poster's essay has been written, English continues to be the language for which most Internet software is developed. The model of the virtual public sphere theoretically grants access to all possible users but de facto limits it to those with certain qualifications.

The issue of differential access to the Internet is central to the dystopian perspective on the impact of the Internet. The core argument here is that the Internet has developed, and continues to evolve, within a global technological-social framework that reflects the interests of different types of privilege: race, class, gender, education, income, ethnicity, urban versus rural, industrially advanced versus developing nations, and the global North versus the South. While not outright dismissive of the political possibilities offered by the Internet for egalitarian modes of social interaction, this reading is highly skeptical of claims made about the Internet's emancipatory restructuring of social space. Sardar's theorization of the Internet as the "cyberspace frontier" is a good example of a work strongly in the dystopian mold.[31] For Sardar, the Internet is the latest expression of the historical desire of the West to colonize non-Western cultures. Sardar argues that having physically colonized most of the non-Western world, the West has now created a virtual world to symbolically reenact that colonization. The cyber-frontier also represents new forms of colonization by the perpetration of acts of epistemological violence on the others of the West. The victims of virtual colonization are the knowledges and histories of non-Western peoples, which can now be "stored and preserved in digital forms," appropriated, and erased when they no longer have any utility.[32]

While Sardar's metaphor of the frontier is useful in drawing attention to a critical question about the appropriation of knowledges and histories through the instruments of technology, its value is ultimately limited. In conceptualizing the frontier as essentially "Western," Sardar does not allow for the fact that within non-Western societies, elites with the privilege of Internet access may operate similarly with respect to disadvantaged groups. In positing an essentialist relationship between the West and "the rest" as one of violence and exploitation, the metaphor by definition cannot recognize other kinds of relationships between the two. Sardar quite rightly points to the problematic nature of celebratory uses of the concept of the frontier in media rhetoric about the Internet.[33] But his own use of the term points to a rejection of technology and modernity per se. The Internet appears to be no more than a dystopic vector deemed intrinsic to the very condition of modernity and technological development.

Between the utopian and the dystopian, we can identify a range of views that may collectively be described as the "entrenched-with-autonomy" model of the Internet. The minimal common assumption that marks different perspectives in this model may be expressed as follows: the uses and understandings of the Internet are mediated through the specificities of the multiple and overlapping contexts—local, national, or global communities; personal, familial, or professional networks; and categories such as class, race, gender, ethnicity, and so on—that users inhabit, but the Internet cannot be reductively viewed as merely coextensive with the patterns of its usage. The Internet is conceptualized as an object whose agency is profoundly implicated in the agency of its users, but, at the same time, one that possesses a material specificity of its own. In terms of the organizing frame of social space, the entrenched-with-autonomy model posits that the space of the Internet is embedded in, interwoven with, and distributed across various other kinds of spaces—for example, economic, technological, cultural, and political. At the same time, the Internet offers unique ways for reconfiguring these other spaces and social space at large.

Castells' influential work defines the Internet as a central feature of the "network society" that has crystallized over the second half of the twentieth century due to innovations in computing and telecommunications, transformations in the workforce and nature of work, and the social impact of the mass media.[34] In the network society, practically every dimension of contemporary society operates in terms of flows of capital, information, technology, and images, sights, and sounds.[35] Spatiality in the network society cannot be understood in terms of traditional conceptions of space but must be grasped as the *space of flows*, the "material organization of time-sharing social practices that work through flows."[36] This material organization is composed of a three-layered network of electronic circuits, strategically significant centers for the smooth functioning of flows, and a spatial grid of transnational elites whose decisions dictate the flows of different resources.[37]

Flows in network society have radically restructured social space locally, nationally, and globally. They have reshaped the geography of large cities where patterns of urban and suburban residence have been altered by a dispersal of key business centers. Flows have given rise to "megacities" across the globe, massively populated locations that attract large migrations of people as centers of economic opportunity.[38] They have created "hinterlands," outposts that service the needs of key centers across the globe and are vitally dependent on these centers. Yet the network society, for all its rapid flows of resources, has created its own forms of inequality. Megacities are both "globally connected and locally disconnected."[39] Elites within these cities are connected through networked flows with elites across the world, while the nonelites are increasingly cut off from even other groups within the cities in which they live. Some parts of the globe are virtually "switched off" from the global network since these regions lack the basic technological capacity to "log on" to the network.

As a technological innovation enabling flows of various kinds, the Internet has significantly contributed to the emergence of the network society. From Castells' description, it is also apparent that the material structure of the Internet approximately mirrors that of the space of flows, since the Internet is comprised of electronic circuits, key hubs of development, and an elite that has access to it. As structurally homologous with the network society, the Internet is an ideal mechanism to support the optimal functioning of network society. The structural limitations of the Internet are those of the network society itself, such as, for instance, the question of differential access.

Castells' notion of the space of flows is a highly sophisticated conceptualization of the Internet and the wider economic, technological, and social contexts within which the Internet is embedded. It does not posit an ontology of the online experience. And it grants the Internet a particularity within the larger apparatus of the network society without dichotomizing the Internet and society as two separate entities. In this model, one's experience of the Internet will depend on the quality of access allowed by the character of one's existence in the network society. Castells' theorization of the space of flows draws attention to the material basis of both spatiality and temporality in society, in its emphasis on "time-sharing" social practices via networks.[40] Incorporating a temporal dimension to the metaphorical understanding of social space shifts the conceptualization of society from object to process. As a *general* mode of thinking social space, the emphasis on temporal flows is valuable, since it does not reify a time-and-place-specific model of society against which the impact of the Internet can be measured as social "change." Yet, Castells' *particular* use of this mode of thinking society—the networked society of flows—also winds up privileging a particular societal model as the norm. Using Western society as the model, Castells proposes a universal logic of development for all societies as moving toward the network society and toward integration within a global network society. This brings with it the danger of misrecognizing as "resistance" those forms of sociality (whether in Western or non-Western societies) that do not conform to this logic. Such forms of sociality and their use of the Internet might possess an autonomous logic that escapes the framework of the network society and the space of flows.

In their important ethnographic analysis of the impact of the Internet on Trinidadian society, Miller and Slater assert that the social uses, meanings, and understandings of the Internet cannot be separated from the concrete realities of the lives of users, whose online activities are in keeping with their goals and interests in different spheres of social life.[41] In keeping with this logic, Trinidadians adopt the Internet as a Trinidadian object, constructing it as "Trini." Trinidadians generally view themselves as modern, international, and cosmopolitan, a fact the authors attribute to the community's long history of emigration to the West. Trinidadians see the Internet as an avenue for the expression of an authentic national identity that is seamlessly integrated within a global whole. The Internet is also viewed as a symbol of values, such as personal freedom and the realization of individual potential.

Yet the use of the Internet has also led to a reshaping of Trinidadians' understanding of their local and global identities, for, logging on, Trinidadians find that although they have always imagined themselves as an international community, they are, however, absent from the imagination of the world. The usual reaction of Trinidadians to this realization is an assertion that the pace of modernization must increase in Trinidad, so that Trinidad can establish a visible presence on the Internet. The government must speed up deregulation, and society must adapt to help Trinidadians capitalize on the potential of the Internet.[42] The anxieties about the Internet are anxieties about modernity itself, but the imagined relationship to modernity is now reoriented around the object of the Internet.

Miller and Slater state that "the object we call the Internet actually consists of groups such as the Trinidadians you will meet in the volume."[43] They advocate a comparative ethnographic approach, comprising of several studies like theirs, across which it may be possible to schematize a general theory of the Internet. They argue for the necessity of implementing this approach according to the "traditional canons of ethnographic inquiry."[44] Their argument implies that their understanding of Internet "space" is firmly based on the understanding of space proposed by the traditional ethnographic approach. But this argument can be critiqued on one ground: the traditional ethnographic method presumes as its model the territorially bounded ethnographic site. Yet, as Miller and Slater point out, "there are huge differences between Trinidadians at home and those who live 'away.'"[45] To account for these differences, the authors combine methods of interviewing and participant observation, spending five weeks in Trinidad, conducting interviews with Trinidadians in New York and London, and gathering information through long-term online relationships with other Trinidadians. But this approach already takes the Internet to be a national object and defines the space of the Internet as a national space, even if, in its inclusion of diasporic communities, it exceeds the limited territorial boundaries of the nation-state. The covert frame and spatial metaphor in the study is that of the nation. A comparative ethnographic approach of Internet usage, as suggested by the authors, might require a more explicit theorization of the transnational space in which Internet usage occurs. It is the wider, transnational, character of cyberspace, for instance, that leads Trinidadians to the realization that Trinidad is relatively marginalized in the online world.

Howard proposes an "embedded media perspective" as a frame for assessing Internet use for a range of social practices and objectives.[46] He locates the Internet among a "range of technologies and technological possibilities labeled *new media*."[47] The paradigm of media embeddedness, through which the dynamics of society online can be grasped, broadly emphasizes a dialectic approach, that is, "how new communication tools are embedded in our lives and how our lives might be embedded in new media."[48] Thanks to the possibilities of interactivity and agency available in new media technologies, these are socially embedded to a far greater degree than traditional communication mechanisms. Users of new media are at once producers and consumers of content; moreover, users can also double up as creators of the apparatus of communication, by designing their own software and hardware. Paradoxically, it is the flexible, fluid, and interactive structure of new media—characteristics lacking in traditional media—that enable new media to be deeply rooted in various aspects of the social lives of users. New media bridge existing and new forms of social practice in that they allow existing practices to be iterated in new modes. For millions of users, new media technologies are an integral part of their lives, whether in the form of new kinds of economic and commercial structures, possibilities for political expression, or unique cultural experiences. New media technologies allow for social practices to be experienced and expressed "in ways that are exciting and new yet deeply ingrained in both the biases and beauties of social life offline."[49] The embedded media perspective also enables an understanding of how people "actually perceive the relationship between new media technology and their quality of life."[50] The fundamental insight offered here is that the dynamics of Internet usage are inextricably linked with users' *perceptions* of the Internet and their experiences of being online. Users of new media technologies believe that their usage of these technologies results in gains in "social capital" or "who we know," as well as "cultural capital" or "what we know."[51] In the perception of users, usage of new media technology is associated with a positive increase in both sociality and knowledge.

The embedded media perspective offers an elegant model of the relationship between the space of the Internet and social space. It specifies the distinct characteristics of new media technologies, as well as indicators that illuminate important aspects of Internet development and usage, within and across communities and contexts. Yet it may not sufficiently take into account the political economy of the production of new media technologies, as a result of which certain key locations across the globe, such as Silicon Valley, are disproportionately powerful in determining developments. Users in developing countries may not have the opportunity to benefit from media technologies that allow them to realize their aspirations, given the lack of the commercial viability of new media products in such markets. As with the notion of the network society, the question here is whether there are forms of sociality in the developing or developed world that are immune to being embedded by new media technologies.

This brief survey helps identify some important issues for a theory of Internet discourse. There is one other important dimension of the relationship between the Internet and society that any such theory must also take into account, specifically, the influence of deeper historical factors that inform Internet usage in particular contexts. For instance, characteristics of online Hindu nationalism are derived from a particular Indian history of valorizing Hinduism as a scientific worldview and reflect an enduring cultural association between national identity and scientificity. The next section outlines a preliminary theoretical model for evaluating the impact of the Internet on social space.

The Cyberfield and Cybercapital: Outline of a Preliminary Model for Analyzing the Impact of the Internet on Social Space

In continuity with the conceptual frame of space, the model for assessing the dynamics of Internet discourse presented here is articulated in terms of Bourdieu's theorization of social space, specifically the concepts of field, capital, and class. The flexibility and generality of Bourdieu's framework allow these concepts to be used in conjunction with elements of other theoretical frameworks. Given the diverse disciplinary nature of the constantly expanding field of Internet studies, such a broad and inclusive framework might be especially relevant for this area of academic inquiry. The domain of Internet development, usage, and discourse can be conceptualized as a field proper in itself that can be termed the *cyberfield*. Correspondingly, capital generated, accumulated, exchanged, and negotiated in the cyberfield can be termed *cybercapital*. While the cyberfield is linked to, and overlaps with, other fields—most notably the technological, but also the economic, cultural, social, and educational—it cannot be subsumed under any of these other fields. As is the case with any field proper, the logic or nomos of the Internet field is not reducible to the logic or nomos of any other field. However, structural factors that have influenced the historical development and continue to inform the circulation of capital in other fields also play a role in orienting the development of the cyberfield. Cybercapital can be defined broadly as the extent to which one may utilize the Internet toward enriching one's quality of life in as many areas as possible. Put another way, cybercapital can be seen as the ability to use the cyberfield to accumulate currency or capital in the maximum number of other fields: for example, economic capital, through using the Internet to enhance productivity and efficiency at work, online shopping, trading online, and the like; social capital, through increasing one's network of social connections; or educational and cultural capital, through increasing one's knowledge, awareness, and so on. As this definition suggests, the possession of Internet-related *professional* skills, such as Web design or software programming, would significantly increase one's cybercapital, through the translatability of such skills into economic capital. In societies like India, the prestige accorded to technological expertise and the current emphasis of state and market on information technology means that such professional qualifications would also endow the possessors of Internet-related skills with significant cultural and symbolic capital. Those individuals who play a role in shaping the development of the Internet and effectively operate as gatekeepers to the Internet—such as Bill Gates, the founder of Microsoft, or Tim Berners-Lee, the inventor of the World Wide Web—would possess very high volumes of cybercapital, economic capital, cultural capital, and symbolic capital. These individuals might also be considered as inhabiting the dominant fraction within a global class of Internet users in a transnational cyberfield, in that they are part of a group that has historically held a privileged position in the field and in that their actions may be interpreted as oriented toward the preservation of that privilege.

These understandings of the cyberfield and cybercapital are consonant with Castells' conceptualization of the Internet as an integral part of the network society and a key component in the transnational space of flows. As Castells argues, a transnational elite that has access to the space of flows controls the circulation of resources in the network society and across its constituent mechanisms such as the Internet. The definitions also allow one to place the development and functioning of the cyberfield in a particular society in the context of a global trajectory of technological development and Internet development. This global trajectory, in turn, reflects the differential levels of economic and political power across societies that are part of a transnational grid of technological development. The definitions of the cyberfield and cybercapital also echo Howard's arguments, already discussed, that people associate Internet usage and familiarity with an increase in their social and cultural capital. As the Internet evolves historically, through both the *practices* and *perceptions* of users, as a means of accumulating different types of capital, the cyberfield continues to gain legitimacy as a field proper and to further entrench its roots across other fields and in society in general. In this regard, Bourdieu's conceptual framework, with its eschewal of the subjective-objective dichotomy, enables us to view users' perceptions or reflexivity about Internet usage as contiguous with their Internet usage itself, without privileging either category over the other.

The relation of the cyberfield to other fields, through the exchange of capital, indicates how the cyberfield may influence changes in other aspects of social existence. It also suggests some modes in which the

Internet may promote the emergence and expression of new forms of social identities. For one, the Internet promotes and accelerates the accumulation of different species of capital among participants in different fields and, in doing so, impacts the differential distribution of capital within and across fields. This map of the spread of species of capital—in particular, symbolic capital—can also be read as a topography of social power relations between different groups of members of a society within and across fields. The Internet itself becomes associated with the possession of social status and symbolic capital.

The modes of communication and expression offered by the Internet can also allow for the relativization or provincialization of truth claims in refashioning or reproducing the online "history" of a community, can lead to the confluence of previously unimagined relationships between discourses as the basis of collective identity, and can foster new logics of personal identity. The example of online Hindu nationalism exemplifies all these processes and possibilities. The double move in online Hindu nationalist discourse of simultaneously insisting on the historical credibility of the Hindutva reading of Indian history while denouncing credible historical work as pseudosecular, doctrinaire Marxist, and anti-Hindu writing is precisely this kind of revisionist narrative grounded in a relativization of the claims of genuine historical scholarship. It bears noting here that technocultural Hindu nationalism forges together capitalism and Hinduism in a manner that would have been alien to Hindu nationalist discourse prior to 1991. While, in a post-1991 scenario, the connection between consumption and Hinduism has also been witnessed in other media, for example, in popular Indian cinema, the Internet seems especially well suited to promoting the idea of Hindu consumption. The linking together of diverse activities under umbrella communicative structures like *Hindu Universe* on the Internet allows for a redefinition of a range of practices as "Hindu."[52] Such associative frameworks that reinforce Hindu identity, combined with the interactivity of Internet communication, provide users with the possibility of performing the act of "Hindu shopping" online in at least two senses—shopping as a Hindu and shopping for Hindu objects. Likewise, buying a cell phone through the *Mobilehindu* Web site takes on the significance of an act of personal self-expression, reflecting a new logic of Hindu identity where one's Hinduness is expressed as much in the purchase of a cell phone as it is in the act of visiting a temple or the recitation of a prayer.[53]

The nomos of the cyberfield, or the stakes according to which the generation of cybercapital takes place, will be unique for each society, reflecting the historical evolution of the cyberfield in that context. The concept of the nomos, in fact, is especially valuable as a heuristic tool in this regard: it necessitates locating the dynamics of Internet discourse and power relations between participants in the cyberfield in any society in terms of the *historicity* of the field in that society. It also follows that the rates of exchange between cybercapital and other kinds of capital will vary from society to society. But we can identify some general factors that would influence the nomos of the cyberfield in every society, although the concrete shape taken by the nomos will always depend on, and will always be embedded in, context-specific conditions.

First, the nomos of the cyberfield in any society depends on the extent to which the Internet is central to technological development in that society. But it should also be clearly noted that Internet development does not exhaust technological development, neither in societies where the cyberfield is already well developed nor in societies where the field is developing or is in nascent form. It is worth repeating here that the cyberfield should not merely be viewed as a component of the technological field and that the two forms of capital are not identical. Rather, they may take on a range of relationships based on the type of society, differentiated in terms of the level of integration of the Internet in technology at large. In postindustrial service economy societies such as the United States—where the Internet is at the forefront of technological development and the Internet is a central part of the technological offerings available to users for a vast range of activities, from communication to education, and from travel to downloading music—cybercapital can easily be cashed in as, or almost fully converted to, technological capital. For example, a self-taught programmer of Internet software languages or an individual who has developed an expertise in strategies for searching the World Wide Web might be employable by technology firms of various kinds, who would find these skills useful. But the corollary that technological capital fully converts to cybercapital in such advanced societies does not necessarily hold true. One can think of the example of a technology professional in the United States who may not necessarily use the Internet for his or her work, beyond communication purposes.

In another kind of society, technological development may cover information technology, to which the Internet is central, as well as the spheres of agriculture and industry, where the Internet may not be as crucial to operations or innovation. For example, in India, state policies and the market now grant information technology and the Internet pride of place in plans of national technological and economic development, but manufacturing and agriculture are not entirely unimportant as focus areas. Cybercapital in India possesses significant exchange value in translating into technological capital as well as economic capital with regard to certain realms of technological activity, such as, for example, developing a Web-based interface for conducting financial transactions. However, there also exist domains of technological activity where cybercapital cannot be traded in for capital. In yet other societies, where the Internet is not critically important to technological development, cybercapital may not offer high returns as technological capital, although it may well translate into other kinds of capital, such as cultural or even symbolic capital, given the prestige associated with access to a service that is not publicly or widely accessible.

A second generalization is that the nomos of the cyberfield in a society will depend on the extent to which there are well-developed equations of exchange between the cyberfield and other fields in that society. In other words, the nomos of the cyberfield will depend on the extent to which the Internet has permeated the various domains of social existence. If the Internet is a vital part of the realm of economic and commercial transactions, that is, the economic field, a lack of familiarity with the Internet will effectively operate as a marker of a lack of ability to generate economic capital. Similarly, in the social field as well as the cultural fields, lack of access to the Internet will operate as—and will be perceived as—social or cultural inferiority, or lack of both types of capital. Concomitantly, in societies where the reach of the Internet is extensive, symbolic capital may also flow from the possession of cybercapital obtained through a variety of means, such as access to high-end equipment or tools that may not be available to the general public; possession of specialized skills and expertise; or holding gatekeeper status in the development of the Internet through a role in government or the private sector. Even in societies where the reach of the Internet is not widespread, Internet usage can generate significant symbolic capital, being perceived as an exclusive activity that is the privilege of an elite. The findings of Miller and Slater's study of Trinidadians' use of the Internet can be interpreted as an affirmation of the Internet by Trinidadians as a source and object of social, cultural, and symbolic capital.

A third factor—and this may distinguish the cyberfield structurally from all other fields, with the possible exception of the technological field—is that the nomos of the cyberfield is not just shaped as part of a *national* history but rather as part of a *global* history. In addition to the policies of the state, in the case of the cyberfield, global forces may directly alter the nomos of the field as well as set the exchange rates between different fields. It is true that the nomos of other fields may also be shaped by external events beyond the control of the nation-state, for example, through the declaration of war on one nation by another. But such situations are exceptional in the case of other fields, such as the cultural, social, political, and so on. The economic field may appear to be another possible exception, for example, given the fluctuations that an increase in the price of oil may cause in a particular national market or economy. But, even in such cases, it is the state that ultimately determines the response of the national economic field to external forces and it is the state that decides whether there should be any dramatic change in the nomos of the field. Now, one can argue that the same observation on the final authority of the state, through control over policy, also holds true for the cyberfield. Indeed, it is not the case that the nation-state is irrelevant or redundant in the case of the cyberfield. Rather, the point to be emphasized here is that the normative, everyday functioning of the cyberfield might be as influenced by global forces as by the agency of the state. It is this fact that structurally distinguishes the cyberfield from other fields.[54] For example, technological applications developed outside the territorial boundaries of the nation-state but downloadable through the Web may transform business practices, communication practices, or cultural practices within a "national" habitus in a short span of time. Instant messengers, Web-based email, search technologies, music file-sharing technologies—all these are examples of developments that have transformed the functioning of the Internet field globally. The extent to which developments related to the Internet result in a radical reconfiguration of the nomos of the cyberfield in a society depends on both the nature of the development and the extent of the reach of the Internet in that society. It is valid to assert, for example, that the invention of the World Wide Web in the early 1990s transformed the nomos of the Internet field *everywhere*.

In distinguishing between types of societies, with respect to the centrality of the Internet in larger structures of technological development or the relative extent of the reach of the Internet, I am not proposing a teleological model of Internet development. My claim is not that all societies will follow the same path and model of Internet development, within the same narrative of technological development or modernization at large. To the contrary, it is quite possible that Internet development will follow different, even unpredictable, lines of evolution in different societies. The extent to which the Internet will converge with, replace, or function in subsidiary status with regard to other technologies—such as the radio, for example—will depend on contextual social, cultural, and economic factors, as well as on the national policies in question. The concepts of cyberfield and cybercapital will also need to be reworked or utilized with other concepts for an effective diagnosis of the complexities of Internet usage and practice in any particular society. With that qualification, I now turn to the concluding section, which employs the model of the cyberfield and cybercapital in offering a comparison of online Hindu nationalism with alternate possible expressions of Indian identity on the Internet. This reflection is interwoven with some thoughts about the general viability of Bourdieu's theoretical framework for assessing the social and cultural impact of the Internet.

Conclusion: Alternative Imaginings of Technology and Indian National Identity on the Internet

Technocultural Hindu nationalism—which typically finds expression on the Internet—emerges historically as a result of the convergence of the doxa of Hindu nationalism and the doxa of neoliberalism, by forging the cultural reason of Hindu nationalism with the economic and meritocratic logic of neoliberalism into a new nomos for the Indian nation-state. Technocultural Hindu nationalism can be viewed as a larger project of altering the nomos of every field in Indian society, but one that, in each domain of Indian social existence, focuses on relevant field-specific issues and manifests itself in field-specific forms. In the cyberfield, the discourse of Hindu nationalists clearly conforms to the fundamental axiom of technocultural Hindu nationalism, that is, the combination of Hindu nationalist cultural argument with support for a strongly neoliberal perspective on economic and technological development. With reference to the dynamics of the cyberfield, this axiom translates into an attempt to redefine Indian cybercapital itself.

The online Hindu nationalists can be viewed as one dominant fraction of the class of participants in the Indian cyberfield, located both within India and in the United States. There may be other segments or fractions of the class that are also struggling for dominance of the field to redefine the nomos of the field. The proposition of the online Hindu nationalists regarding the nomos of the Indian cyberfield can be described as follows: the act of using the Internet is an activity that is simultaneously Hindu, Indian, and global, but of these three elements, it is the first that is most important and constitutive of the others. It is "Hinduness" that forms the basis of one's Indianness as well as the substance of one's globality. It is the Hindu quality of innate technological ability from which a national Indian genius in technology is derived, and it is this intrinsic Hindu quality that also empowers one to be an ideal global citizen and global consumer. The Hindu nationalist aspiration for altering the nomos of the cyberfield also involves a conflation of technological practice and cultural practice as an expression of one's cultural-national identity. To be technologically adept is to be a good Hindu, and vice versa. If the Internet is a shining symbol of the brave new technologically driven globalized world, so too is Hinduism, with its eternally valid axioms, its universalism, and its indubitable scientificity.

As Rajagopal has noted, Hindu nationalists see the Internet "not simply as a channel of communication, but as a key battleground that must be occupied and defended."[55] The militaristic metaphor succinctly captures the Hindu nationalist ambition to inscribe the space of the "Indian" Internet with a normative logic of what it means to be Hindu-Indian. The Hindu nationalist propagation of a discourse of Hindu-Indian identity (that is at once a narrative of authenticity) toward colonizing Indian cyberspace operates through a variety of strategies online. I have already discussed the techniques employed in online Hindu nationalism to promote the idea of Hindu consumption or the notion of consumption itself as a Hindu practice. These techniques are supplemented online by a range of hostile strategies adopted by the self-appointed Hindu nationalist cyber-guardians of Hinduism and Indianness. For example, the *Hindu Unity* Web site boasts a "black list" or "hit list" of so-called anti-Hindus, which includes Pervez Musharraf, the president of Pakistan; Pat Robertson, the American evangelist; Indian historian Romila Thapar; and several journalists.[56] Hindu nationalists have also used the Internet to target and intimidate scholars perceived to be "anti-Hindu."[57] Not explicitly militant but no less prescriptive are rigid and essentialist constructions of Hinduism online, as seen, for example, in the section titled "On the Nine Beliefs of Hinduism" on the *Himalayan Academy* Web site.[58]

What this points to is the urgent necessity for developing a counter-hegemonic discourse to Hindu nationalism in cyberspace, one that can achieve the same level of visibility online and can as effectively use the Internet as a tool of communication, organization, and empowerment in promoting alternate notions of Indian identity. This is not to imply that other forms of Hindu or Indian identity are lacking a presence on the Internet. For example, *Communalism Combat*, *Progressive South Asians*, the *SARAI* project, and *South Asian Citizen's Web* are virtual civil society initiatives committed to fashioning alternative articulations of Indian identity online; to promoting values of human rights, secularism, and democracy; and to creatively using the Internet as a space for building community.[59] But the centrality of the Internet to the propagation of Hindu nationalist discourse seems distinctive in comparison to the use of the Internet by other Indian nationalist, religious-cultural, or secular discourses. Indian communities are not lacking for representation on the Internet either: numerous Indian religious, regional, linguistic, ethnic, and caste groups have their own news Web sites, discussion forums, and portals on the Internet. However, the discourse of Indian identity on the Internet is essentially fragmented along the axes of particularistic identities, including the categories of secular Indian identity or South Asian civilizational identity, which appear as just two other visions of Indianness.

But this raises complicated questions. Is it possible or desirable to envision some minimal element around which to orient and construct Indian identity *online*, given the possibly undemocratic and prescriptive nature of such an exercise? One might think, rather, of an ethics of technological practice as an alternative performance of Indianness in cyberspace as the minimal condition for developing a counterhegemonic discourse. What may separate the technocultural Hindu nationalists from other classes of Indians in the cyberfield is the strong emphasis on the part of the former on *understanding* and *using* technology as a medium and instrument for promoting their discourse. Without necessarily agreeing with the propositions of the neoliberal paradigm or, for that matter, of Hindu nationalist ideology, Indian individuals and groups committed to fostering more inclusive notions of Indian identities may need to adopt a similarly proactive approach toward understanding and using technology.

To make this argument is not to be in agreement about the inevitability of neoliberalism as a national or global paradigm of development and governance; rather, it is to see larger processes such as globalization and object spaces such as the Internet in terms of a complex amalgam of possibilities and constraints, a mix of opportunities as well as challenges. As the Indian state continues to gradually recalibrate its policies toward increasing global technological and economic integration, it will increasingly validate those expressions of Indian identity—especially in its policies regarding technological development or the development of the Internet—that appear ideally suited for functioning within such an integrated national-global frame. But this may not necessarily be to the exclusion or detriment of other inclusive articulations of Indian identity, whether online or offline. It is interesting to note, for example, that the *Indian Diaspora* Web site described in the first section of this chapter does not praise Hindu nationalism as the basis of Indian achievement, but rather lauds *Indian technological* and *professional* ability as the source of diasporic Indian success.

Can one, then, imagine ways in which inclusive visions of Indianness can be linked through the space of the Internet (among other forums and spaces) in a meaningful way? The point of this speculation is not to propose a prescriptive or authoritative idea of Indian identity online that is applicable in simplistic or standardized fashion to all Indians who use or hope to use the Internet. The goal, much more modest, is to try and think of some attributes of a cyberfield that can operate as a viable alternative to technocultural Hindu nationalism, in the knowledge and awareness that Internet access in India is limited to a fraction of the population. It may be apposite here, in the context of such an exercise, to address some limitations of the theoretical framework used in this book for tracing the relationship between technology and nationalism in India and for describing technocultural Hindu nationalism. For it is by recognizing the limitations of the framework that we can adequately make use of its strengths and potential. Bourdieu's theoretical framework may be critiqued on the same general ground on which Gandhi critiqued modernity itself. That objection, simply stated, is whether aspects of human social life can be completely described in terms of the calculations of capital. From the perspective of the Gandhian critique of modernity, in so far as the vocabulary of field, class, capital, and the theory of the state suggests that this is indeed the case, it treats as a given that we dwell in the condition of modernity.

Now, as explained in chapter 2, Bourdieu does not conflate the concepts with the objects of research. His claim is not that a person, group, or society *actually* functions in this or that manner. The claim, rather, is that a person, group, or society can be viewed *as if* they were functioning in accordance with certain theoretical concepts. However, it may be the case that, in the Indian context or elsewhere, there are forms of sociality that resist being described in terms of the exchange of capital.[60] And Bourdieu's framework may not be effective in identifying or enabling an analysis of these forms of sociality. Some implications of this fact about Bourdieu's theoretical framework may be noted. The framework may misrecognize or refuse to recognize the existence of alternative modes of Indian social life that continue to operate beside (and as a bulwark against the rise of) Hindu nationalism and yet do not necessarily fit the paradigm of secularism associated with Indian constitutional democracy and political modernity.[61] In making this observation, I am not suggesting that there is a realm of authentic Indian social practices that escapes being detected by a particular epistemological framework such as Bourdieu's. My point is simply to highlight that there may exist some possibilities for thinking social practices, including technological use, outside the play of one form of capital or another, or outside the metaphors of games. These possibilities can accompany, complement, and enrich an analysis that draws on Bourdieu's framework. At the same time, the reliance of Bourdieu's framework on the priority of modern forms of social existence over others may, arguably, have a value, even if unintended, in the Indian context. The framework operates as a critique of the essentialist and primordial claims of Hindu nationalists about the scientific and universalist nature of Hinduism or about the superior disposition of Hindus that empowers them to participate in a technologically powered global capitalist modernity. The framework also complicates the distinctions between the Nehruvian and Hindu nationalist understandings of postcolonial and global modernity and the value of technology by showing these understandings as related to social practices and produced by the play of capital. And, perhaps most importantly, in showing us that no social arrangement or state of affairs is inevitable or necessary, it holds out the possibility of imagining and achieving modes of social practice that are structured by principles that we recognize as more egalitarian than our current modes of sociality.

I conclude, then, with a hopeful, if utopian, gesture toward the possibility of inclusive forms and modes of social and technological practice on the Internet and elsewhere. The long and complex historical association between technology and nationalism in India has taken expression through many forms, for example, in the avatar of state policies and social scripts that have been empowering and disempowering, enabling as well as restrictive. The relationship between Indian technology and nationalism has borne the imprimatur of notions of Indian selfhood and modernity as much as it has constituted these Indian selves and modernities. The relationship also forms the historical basis for the peculiar commingling of Hindu nationalism and cyber-technologies. But the relationship may also engender existing, incipient, or as yet unimagined progressive visions of Indian identity—visions that need to be made visible and endowed with legitimacy, in cyberspace and elsewhere, among inhabiting Indians and the diaspora.

Endnotes

1. Paul du Gay et al. identify representation, identity, production, consumption, and regulation as five elements or aspects of a "circuit of culture…through which any analysis of a cultural text or artefact must pass if it is to be adequately studied." Paul du Gay, Stuart Hall, Linda Janes, Hugh Mackay, and Keith Negus, *Doing Cultural Studies: The Story of the Sony Walkman* (London: The Open University, 1997), 3. The model may be helpful in illuminating some of the dynamics of usage of a media and technological form like the Internet, but it may need to be radically revised if it is to be applied as a general explanatory framework for analyzing online discourse. The use of the Internet by communities, for instance, does not inevitably parse into the five distinct stages of the circuit of culture, and additional explicatory or descriptive categories may be required for specific kinds of uses of the Internet by online groups. The stages of production and consumption may in some cases overlap online. Circulation, a category not included in the circuit of culture, may be central to the sense by which participants in online discourse perform their allegiance to the obligations of an imagined community. Also, compared to the Internet, the Sony Walkman—the particular object analyzed in the case study by du Gay et al.—is a relatively discrete-bounded entity whose existence in different arenas of social life may be somewhat easier to trace in terms of "the social identities associated with it" as compared to the Internet (3). To be sure, the social identities that may be associated with the Walkman as well as the Internet are, respectively, contingent on numerous factors. But the Internet as a new and unconventional media form, which, among other things, is itself a site of the production of discourses of social identity and may require forms of analysis that go beyond the materialist mode of analysis recommended by du Gay et al.
2. In this chapter, I use online Hindu nationalism as synonymous with techno-cultural Hindu nationalism.
3. Government of India, Ministry of External Affairs, Non Resident Indians & Persons of Indian Origin Division, *The Indian Diaspora* (last updated October 5, 2004), http://indiandiaspora.nic.in (accessed February 11, 2005).
4. See idem, "Mandate of the Committee," *The Indian Diaspora*, http://indiandiaspora.nic.in/mandate.htm (accessed February 11, 2005); and idem, *Report of the High Level Committee on the Indian Diaspora* (December

2001), http://indiandiaspora.nic.in/diasporapdf/part1-est.pdf (accessed February 11, 2005). Hereafter *HLC Report.*
5. Idem, *HLC Report*, "Foreword," v–x, http://indiandiaspora.nic.in/diasporapdf/part1-for.pdf (accessed February 11, 2005).
6. Ibid., viii, emphasis added.
7. Ibid., vi.
8. Ibid., vi.
9. Ibid., vi.
10. Ibid., vi.
11. Ibid., vi.
12. Ibid., vii.
13. The substantive definitions of Indianness are different in Hindu nationalist Web sites and the *HLC Report*, but the underlying claim of *historical credibility* is common to both.
14. Anne O. Krueger and Sajjid Chinoy, "The Indian Economy in Global Context," 9–45.
15. Manuel Castells, *The Internet Galaxy: Reflections on the Internet, Business and Society* (Oxford: Oxford University Press, 2001).
16. Ibid., 118.
17. Ibid., 118.
18. Bosah Ebo, "Internet or Outernet?" 1–12.
19. Ibid., 2–3.
20. Ibid., 5–6.
21. Howard Rheingold, *The Virtual Community.*
22. Ibid., xviii.
23. Ibid., 10.
24. Mark Poster, "Cyberdemocracy: The Internet and the Public Sphere," 212–228.
25. Ibid., 222.
26. Ibid., 222.
27. Ibid., 225.
28. Ibid., 224.
29. Ibid., 218.
30. This chapter refers to the revised 2000 edition of the book. See Rheingold, *The Virtual Community*, 323, 345.
31. Ziaudddin Sardar, "alt.civilizations.faq," 18.
32. Ibid., 19.
33. Ibid., 17.

34. Manuel Castells, *The Rise of the Network Society*, 2nd ed. (Oxford: Blackwell, 2000).
35. Ibid., 442.
36. Ibid., 442.
37. Ibid., 443.
38. Ibid., 443.
39. Ibid., 436.
40. Ibid., 442.
41. Daniel Miller and Don Slater, *The Internet: An Ethnographic Approach* (Oxford, U.K.: Berg, 2000).
42. Ibid., 17.
43. Ibid., 8.
44. Ibid., 21.
45. Ibid., 3.
46. See Philip N. Howard, "Embedded Media: Who We Know, What We Know, and Society Online," in *Society Online: The Internet in Context*, ed. Philip. N. Howard and Steve Jones (Thousand Oaks, CA: Sage Publications, 2004), 1–27.
47. Ibid., 1, emphasis in original.
48. Ibid., 2.
49. Ibid., 22.
50. Ibid., 19.
51. Ibid., 14.
52. *Hindu Universe* (accessed February 19, 2005).
53. *Mobilehindu* (accessed February 11, 2005).
54. The technological field, to some extent, may also share this characteristic of the cyberfield.
55. Rajagopal, "Hindu Nationalism in the US," 485.
56. "Hindu Unity's Criminals Hit List: The Black List, Enemies of Hinduism Exposed," *Hindu Unity*, http://hinduunity.org/hitlist.html (accessed March 21, 2006).
57. See, for instance, Shankar Vedantam, "Wrath Over a Hindu God: U.S. Scholars' Writings Draw Threats From Faithful," *Washington Post*, April 10, 2004, A01: http://www.washingtonpost.com/ac2/wp-dyn?pagename=article&contentId=A334-2004Apr9&notFound=true (accessed August 11, 2005).
58. "Nine Beliefs of Hinduism," *Himalayan Academy*, http://www.himalayanacademy.com/basics/nineb/index.html (accessed February 11, 2005).

59. *Communalism Combat*, http://www.sabrang.com/ (accessed August 10, 2005); *Progressive South Asians*, http://www.proxsa.org/ (accessed August 10, 2005); *SARAI*, http://www.sarai.net/ (accessed August 10, 2005); *South Asia Citizens Web*, http://www.sacw.net/ (accessed August 10, 2005).
60. This objection is somewhat different from the charge of instrumentality leveled against Bourdieu's theoretical framework. The charge of instrumentality does not rest on the modern or nonmodern character of the social practices in question. See chapter 2 for a discussion of this issue.
61. Ashis Nandy has argued that Indian social life is not exhausted by the modern, in either its "progressive" or chauvinist variants, which he sees as sharing certain fundamental characteristics. See, for instance, Ashis Nandy, *Time Warps: Silent and Evasive Pasts in Indian Politics and Religion* (New Brunswick: Rutgers University Press, 2002).

Appendix A

Unique Primary Resources of Hindu Nationalism Online

Table A1 lists all the unique primary resources of Hindu nationalism online identified from the results of the five keyword searches conducted between April 5, 2005, and April 8, 2005. Repeats within and across search categories have not been counted. For the name and descriptor of the resource, either the phrase displayed on a masthead or the phrase displayed on the browser title bar of the main home page has been used.

Table A1. List of unique primary resources of Hindu nationalism online.

No.	Name and Descriptor of Primary Resource	URL
1.	Hindutva	www.hindutva.org
2.	Hindutva Series: Hindu History—A Search for our Present in History	http://newsanalysis-intl.tripod.com/
3.	Hindu Unity—Soldiers of Hindutva! Awake Hindus!!	http://www.hinduunity.org
4.	The Hindu Universe—Hindu Resource Center	http://www.hindunet.org
5.	Hindutva	http://www.geocities.com/jairama1/
6.	Books on Hindutva, Hinduism and Nationalism, Minority in India, Saffron Ideology, Prafull Gorad	http://www.prafullgoradia.com
7.	Yahoo Groups—savarkariyahindutva	http://groups.yahoo.com/group/savarkariyahindutva/
8.	The Arun Shourie Site	http://arunshourie.voiceofdharma.com
9.	BJP: Bharatiya Janata Party, India at www.bjp.org	http://www.bjp.org
10.	HinduLinks Universe	http://www.hindulinks.org
11.	Hindu Vivek Kendra	http://www.hvk.org
12.	Hindu Swayamsevak Sangh	www.hssworld.org

(*continued on next page*)

TABLE A1. *(continued)*

No.	Name and Descriptor of Primary Resource	URL
13.	Welcome to Koenraad Elst's Indology Site	http://koenraadelst.bharatvani.org
14.	Hinduism Today Magazine	http://www.hinduismtoday.com
15.	Voice of Dharma	http://www.voi.org
16.	A Tribute to Hinduism	http://www.atributetohinduism.com
17.	Kauai's Hindu Monastery Web Site	http://www.himalayanacademy.com/
18.	Ancient India's Contribution to Science and Technology	http://india.coolatlanta.com/GreatPages/sudheer/
19.	hindu world—main	http://www.hindu-world.com
20.	Sword of Truth Cover Page	http://www.swordoftruth.com
21.	VandeMataram.com—Learn about Indian Civilization, Culture, Yoga, Ayurveda, History	http://www.vandemataram.com
22.	Hindu Student Council @ UCI	http://www.hscuci.org/main.php
23.	HSC Home Page, Univ Illinois at Chicago	http://www2.uic.edu/stud_orgs/religion/hindu/index2.html
24.	University of Florida HSC	http://grove.ufl.edu/~hsc/index.htm
25.	Kashmir: A Paradise Turned into Hell by Terrorism	http://www.kashmir-information.com
26.	Hindu Students Council	http://www.hscnet.org
27.	FreeIndia.org—India site dedicated to freedom movement, education, culture	http://www.freeindia.org/dynamic/
28.	Organiser	http://www.organiser.org/dynamic/
29.	Bhartiya Janta Party—Gujarat	http://www.bjpguj.org/
30.	Kashmir News Network	http://ikashmir.org/new/ikashmir.php
31.	Sewa International	http://www.sewainternational.org/dynamic
32.	SEVABHARATHI.ORG	http://www.sevabharathi.org/
33.	India Development And Relief Fund	www.idrf.org/dynamic
34.	Hindu Web Universe: Free Hindu Web Pages and Services	http://www.hinduweb.org
35.	VNN—Vaishnava News Network	http://www.vnn.org
36.	Vishwa Hindu Parishad of America	http://www.vhp-america.org/dynamic
	Total number of unique primary resources = 36	

Appendix B

Unique Primary Resources of Sikh Nationalism Online

Table B1 lists all the unique primary resources of Sikh nationalism online identified from the results of the searches conducted on April 7, 2005, and April 8, 2005. Repeats within and across search categories have not been counted. For the name and descriptor of the resource, either the phrase displayed on a masthead or the phrase displayed on the browser title bar of the main home page has been used.

Table B1. List of unique primary resources of Sikh nationalism online.

No.	Name and Descriptor of Primary Resource	URL
1.	Khalistan—The New Global Reality	http://www.khalistan.net/
2.	Council of Khalistan	http://www.khalistan.com/
3.	Khalistan Affairs Center	http://www.khalistan-affairs.org/
4.	Khalistani—Sikhs, UK, Pakistan, india, khalsa, lion, swords, war,...	http://www.khalistan.8m.com/
5.	Search Sikhism home page	http://www.searchsikhism.com
6.	S I K H L I O N Z.C O M Raj Karega Khalsa Sikh Lionz Khalistani	http://www.sikhlionz.com/lionzsection1.htm
7.	The Pathic Weekly	http://www.panthic.org
8.	The Khalsa Day Parade: April 10, 2004: Khalistan	http://khalsadayparade.org/index2.htm
9.	Khalistanindex	http://indianterrorism.bravepages.com/khalistanindex.htm
10.	FORT: Panth Khalsa	http://www.panthkhalsa.org/
11.	Welcome to website about history of the sikhs	http://www.sikh-history.com

(*continued on next page*)

TABLE B1. *(continued)*

No.	Name and Descriptor of Primary Resource	URL
12.	Sikhtoons—Sikh cartoons with a message	http://www.sikhtoons.com
—	Resource inactive on April 13, 2005, and May 15, 2005	http://www.surbut-khalsa.com/01.general%20info/new%20index%20contents.htm
13.	OSSA—Ontario Sikh Students Association	http://www.interlog.com/~sikhs/ossa/
14.	Dal Khalsa	http://www.dalkhalsa.com/
15.	Abuse of Human Rights of Sikhs in India	http://www.maboli.com/Sikh_HR
	Total number of unique primary resources = 15	

Appendix C

Unique Primary Resources of Kashmiri Nationalism Online

Table C1 lists all the unique primary resources of Kashmiri nationalism online identified from the results of the keyword searches conducted on April 11, 2005, and April 12, 2005. Repeats within and across search categories have not been counted. For the name and descriptor of the resource, either the phrase displayed on a masthead or the phrase displayed on the browser title bar of the main home page has been used.

Table C1. List of unique primary resources of Kashmiri nationalism online.

No.	Name and Descriptor of Primary Resource	URL
1.	Jammu & Kashmir	http://www.ummah.org.uk/kashmir/
2.	Jammu-Kashmir.Com Homepage	http://www.jammu-kashmir.com/index.html
3.	Kashmir Home Page	http://alumnus.caltech.edu/~mughal/kashmir/kashmir.html
4.	Untitled Document / Friends of Kashmir	http://www.kashmir.co.uk/
5.	Kashmir Media Service	http://www.kmsnews.org/
6.	Jammu Kashmir Liberation Front—JKLF	http://www.geocities.com/jklf-kashmir/
7.	Kashmir—Paradise on Fire	http://www.geocities.com/CapitolHill/3023/kashmir.html
8.	Kashmir Point, Hong Kong	http://www.kashmirpoint.com/
9.	kashmir, kashmir news etc / Kashmir Independence Organization	http://www.angelfire.com/bc/cotr/kashmir1.html
10.	Kashmir Council of Australia	http://home.iprimus.com.au/mianm/kca/index.htm
11.	Holocaust in Kashmir	http://www.holocaustinkashmir.50megs.com/
12.	JKLF's webpage	http://shell.comsats.net.pk/~jklf/

(continued on next page)

TABLE C1. *(continued)*

No.	Name and Descriptor of Primary Resource	URL
13.	Factsheet / Kashmir	http://www.infopak.gov.pk/public/kashmir/kashmir.htm
14.	Jamaat-e-Islami Pakistan (Pakistan India Relations and Kashmir Dispute section main page)	http://www.jamaat.org/issues/kashmir.html
15.	News Desk—Kashmir/Pakistan Awami Tehreek	http://www.pat-pk.com/newsdesk/kasmir.shtml
16.	Yahoo Groups: pakgovnet-kashmir—Kashmir Information Network	http://groups.yahoo.com/group/pakgovnet-kashmir/
17.	Balawaristan National Front	http://balawaristan.net/english.htm
18.	Kashmir Watch: Daily News	http://www.kashmirwatch.com/
19.	Jammu Kashmir Liberation Fron [*sic*] (JKLF_UK_Europe)	http://www.geocities.com/CapitolHill/Lobby/8125
—	Not accessible as of April 28, 2005, and May 15, 2005	http://www.pakwatan.com/main/kashmir/
20.	Azad Jammu and Kashmir	http://www.ajk.gov.pk/site/index.php
21.	HEZB-E-MUGHALSTAN Homepage—The "Party of Mughalstan"	http://www.dalitstan.org/mughalstan/azad/whatisms.html
	Total number of unique primary resources = 21	

Appendix D

Primary Web Resource Pages Representing Units of Online Nationalisms

Tables D1, D2, and D3 list the Web pages from the resources that were saved to a disk on May 5, 2005, for the purpose of analysis. Pages that did not save or were overlooked by error were downloaded or browsed for directly obtaining text between May 6, 2005, and May 9, 2005.

Table D1. Primary Web resource pages for Hindu nationalism online.

No.	Primary Resource (Shortened Name)	Text Selected/ Constructed as Unit of Nationalism	URL
1.	Hindutva	Introductory text and motto from home page	www.hindutva.org
2.	Hindu History	Introductory text, motto, and what is Hindutva from Home page	http://newsanalysis-intl.tripod.com/
3.	Hindu Unity	Home page content, including descriptor and quotes	http://www.hinduunity.org
4.	The Hindu Universe	More Information/ About/Contact	http://www.hindunet.com/contact.htm
5.	Hindutva	Purpose	http://www.geocities.com/jairama1/purpose.htm
6.	Prafull Goradia	About the Author	http://www.prafullgoradia.com/author.html
7.	Savarkariyahindutva	Home page content, description, and motto	http://groups.yahoo.com/group/savarkariyahindutva/
8.	Arun Shourie Site	Arun Shourie on Hindutva section, main page text	http://arunshourie.voiceofdharma.com/hindutva.htm

(*continued on next page*)

TABLE D1. *(continued)*

No.	Primary Resource (Shortened Name)	Text Selected/ Constructed as Unit of Nationalism	URL
8.	Arun Shourie Site	Arun Shourie on Hindutva section, main page text	http://arunshourie.voiceofdharma.com/hindutva.htm
9.	BJP	Philosophy section—Hindutva link—Hindutva the great nationalist ideology article	http://www.bjp.org/history/hindu.htm
10.	HinduLinks	More Information/About/Contact	http://www.hindunet.com/contact.htm
11.	Hindu Vivek Kendra	About HVK	http://www.hvk.org/abouthvk.html
12.	Hindu Swayamsevak Sangh U.S.—About us	U.S. site About us	http://www.hssus.org/content/view/56/83/
13.	Koenraad Elst	About	http://koenraadelst.bharatvani.org/about.html
14.	Hinduism Today	About Us	http://www.hinduismtoday.com/about_us.shtml
15.	Voice of Dharma	Home	http://www.voi.org
16.	A Tribute to Hinduism	First page with content	http://www.atributetohinduism.com/index_old.htm
17.	Hindu Monastery	About Us	http://www.himalayanacademy.com/ssc/
18.	Ancient India—S/T	Home page	http://india.coolatlanta.com/GreatPages/sudheer/
19.	hindu world	Home page	http://www.hindu-world.com
20.	Sword of Truth	About Us	http://www.swordoftruth.com/swordoftruth/aboutus/aboutus.html
21.	VandeMataram.com	About Us—Our Mission	http://www.vandemataram.com/html/ourvoice/ov-mission.htm
22.	Hindu Student Council @ UCI	About	http://www.hscuci.org/about.html
23.	HSC Home Page (Univ of Illinois at Chicago)	Home	http://www2.uic.edu/stud_orgs/religion/hindu/index2.html
24.	University of Florida HSC	About	http://grove.ufl.edu/~hsc/hsc.htm

(continued on next page)

TABLE D1. *(continued)*

No.	Primary Resource (Shortened Name)	Text Selected/ Constructed as Unit of Nationalism	URL
25.	Kashmir Information Network	Home	http://www.kashmir-information.com
26.	Hindu Students Council	About Us	http://www.hscnet.org/about.php
27.	FreeIndia.org	More Info/ Contact/ About	http://www.hindunet.com/contact.htm
28.	Organiser	About Us	http://www.organiser.org/dynamic/modules.php?name=Content&pa=showpage&pid=6
29.	Bharatiya Janata Party—Gujarat	Philosophy (only first page out of 13 pages)	http://www.bjpguj.org/philosophy.htm
30.	Kashmir News Network	Home	http://ikashmir.org/new/ikashmir.php
31.	Sewa International	About Us	http://www.sewainternational.org/dynamic/modules.php?name=Content&pa=showpage&pid=129
32.	SEVABHARATHI.ORG	Annual report 2004/ About Sevabharathi	http://www.sevabharathi.org/report2004.htm
33.	India Development And Relief Fund	About Us	http://www.idrf.org/dynamic/modules.php?name=Hncontent&pa=showpage&pid=2
34.	Hindu Web Universe	Contact/About/ More Info	http://www.hindunet.com/contact.htm
35.	Vaishnava News Network	About	http://www.vnn.org/about/index.html
36.	Vishwa Hindu Parishad of America	About Us	http://www.vhp-america.org/aboutus/

TABLE D2. Primary Web resource pages for Sikh nationalism online.

No.	Primary Resource (Shortened Name)	Text Selected/ Constructed as Unit of Nationalism	URL
1.	Khalistan	Home page	http://www.khalistan.net/
2.	Council of Khalistan	Open Letter to the Sikh Nation	http://www.khalistan.com/OpenLetters/ol040505_VaisakhiDayMessage2005.html
3.	Khalistan Affairs Center	Khalistan Calling main page as of May 4, 2005	http://www.khalistan-affairs.org/Main/K_Calling/kc05042005.htm
4.	Khalistani—Sikhs, UK etc	Home	http://www.khalistan.8m.com/
5.	Search Sikhism home page	Home	http://www.searchsikhism.com
6.	S I K H L I O N Z.C O M	About Sikh Lionz	http://www.sikhlionz.com/lionzsection1.htm
7.	The Panthic Weekly	About Us section on Waheguroo—parent/umbrella organization	http://help.waheguroo.com/?site=About
8.	The Khalsa Day Parade: April 10, 2004: Khalistan	Our Mission	http://khalsadayparade.org/mission.htm
9.	khalistanindex	Council of Khalistan's open letter to the Sikh nation presented on the site	http://indianterrorism.bravepages.com/Open_Letter_to_\Sikh_Nation.htm
10.	FORT: Panth Khalsa	About this Site	http://www.panthkhalsa.org/about_fpk.php
11.	Hindu Vivek Kendra	About HVK	http://www.hvk.org/abouthvk.html
13.	OSSA—Ontario Sikh Students Association	Home	http://www.interlog.com/~sikhs/ossa/
14.	Dal Khalsa	Full report	http://www.dalkhalsa.com/26jan05/fullreport.html
15.	Abuse of Human Rights of Sikhs in India	Reasons page	http://www.maboli.com/Sikh_HR/pages/first.htmld/index.html

TABLE D3. Primary Web resource pages for Kashmiri nationalism online.

No.	Primary Resource (Shortened Name)	Text Selected/ Constructed as Unit of Nationalism	URL
1.	Jammu & Kashmir	Home page	http://www.ummah.org.uk/kashmir/
2.	Jammu-Kashmir.Com	About	http://www.jammu-kashmir.com/aboutus/aboutus.html
3.	Kashmir Home Page	Home page	http://alumnus.caltech.edu/~mughal/kashmir/kashmir.html
4.	Untitled Document/ Friends of Kashmir	Contact Us	http://www.kashmir.co.uk/fok.htm
5.	Kashmir Media Service	About Us	http://www.kmsnews.org/HTML/about.htm
6.	Jammu Kashmir Liberation Front—JKLF	Mission	http://www.geocities.com/jklf-kashmir/jklfmission.html
7.	Kashmir-Paradise on Fire	Home page	http://www.geocities.com/CapitolHill/3023/kashmir. html
8.	Kashmir Point, Hong Kong	About Us	http://www.kashmirpoint.com/aboutkashmir. shtml
9.	kashmir, kashmir news etc/ Kashmir Independence Organization	About Us	http://www.angelfire.com/bc/cotr/disclaimer.html
10.	Kashmir Council of Australia	About Us—Objectives	http://home.iprimus. com.au/mianm/kca/ index.htm
11.	Holocaust in Kashmir	Home page	http://www.holocaustinkashmir.50megs.com/
12.	JKLF's webpage	Our Ideology, Aims, and Objectives	http://shell.comsats. net.pk/~jklf/i2.htm
13.	Factsheet/Kashmir	Factsheet	http://www.infopak.gov.pk/public/kashmir/facts-kashmir.htm
14.	Jamaat-e-Islami Pakistan (Pakistan India Relations and Kashmir Dispute section)	Pakistan-India Relations and Kashmir Dispute main page	http://www.jamaat.org/issues/kashmir.html
15.	News Desk—Kashmir/ Pakistan Awami Tehreek	Home	http://www.pat-pk.com/newsdesk/ kasmir.shtml
16.	Yahoo Groups: pakgovnet-kashmir—Kashmir Information Network	Home	http://groups.yahoo.com/group/pakgovnet-kashmir/

(*continued on next page*)

TABLE D3. *(continued)*

No.	Primary Resource (Shortened Name)	Text Selected/ Constructed as Unit of Nationalism	URL
17.	Balawaristan National Front	Party Profile	http://balawaristan.net/party.htm
18.	Kashmir Watch: Daily News	About Us	http://www.kashmirwatch.com/feedback.htm
19.	Jammu Kashmir Liberation Fron [*sic*] (JKLF_UK_Europe)	Mission statement	http://www.geocities.com/CapitolHill/Lobby/ 8125/jklfmission.html
20.	Azad Jammu and Kashmir	Conflict	http://www.ajk.gov.pk/site/index.php?option=com_content&task=category& sectionid=34&id=186& Itemid=126#up
21.	HEZB-E-MUGHALSTAN Homepage—The "Party of Mughalstan"	What is Mughalstan?	http://www.dalitstan.org/mughalstan/azad/whatisms.html

Appendix E

Unweighted Frequency Count for Relevant Occurrences of Subthemes and Themes

Table E1. Unweighted frequency counts for subthemes and themes.

No.	Theme	Subtheme	Hindu Nationalism	Sikh Nationalism	Kashmiri Nationalism
1.	Globality (G)				
		Global	15	3	7
		Globality	0	0	0
		Globalization/ Globalisation	1	0	0
		Globalized/ Globalised	0	0	0
		Globally	1	0	0
		Globalism	1	0	0
		Globe	7	1	0
		International	18	1	31
		Internationally	1	0	2
		Internationality	0	0	0
		World/World's/ Worlds/ World-wide/ Worldwide	124	21	34
	Theme G total		168	26	74
2.	Rights and Freedoms (RF)				
		Right	2	2	26
		Rights and-rights (as in human-rights)	5	16	23
		Free	5	10	11
		Freed	1	0	1
		Freely	2	0	2
		Freedom	37	33	26

(*continued on next page*)

TABLE E1. *(continued)*

No.	Theme	Subtheme	Hindu Nationalism	Sikh Nationalism	Kashmiri Nationalism
		Freedoms	0	1	0
		Liberty	4	1	0
		Liberties	0	0	0
		Liberation	0	4	5
		Self-determination	0	1	13
		Sovereignty	0	17	4
		Sovereign	0	6	2
		Sovereigns	0	0	0
		Democracy	4	2	6
		Democracies	0	0	0
		Democratic	1	7	5
		Democratising/ Democratizing	0	0	0
		Independence	9	9	19
		Independent	2	11	6
		Violation	0	0	1
		Violations	0	4	4
		Violating	0	1	2
		Violated	0	1	0
	Theme RF total		72	126	156
3.	Holocaust and Genocide (HG)				
		Holocaust	3	2	5
		Genocide	0	3	1
		Genocidal	0	0	0
	Theme HG total		3	5	6
4.	Terror and Violence (TV)				
		Terror	2	2	2
		Terrorism	4	7	2
		Terrorist	1	5	2
		Terrorists	4	3	0
		Terrorized	1	0	0
		Violence	2	4	0
		Violent	3	0	1
		Non-violence	6	0	0
		Non-violent	3	0	0
		War	7	0	16
		Wars	0	0	3

(continued on next page)

Table E1. *(continued)*

No.	Theme	Subtheme	Hindu Nationalism	Sikh Nationalism	Kashmiri Nationalism
		Warfare	0	0	1
		Kill	1	3	1
		Killed	3	4	6
		Killing	2	0	5
		Killings	0	1	3
		Death	4	4	2
		Deaths	0	1	1
		Prison	0	1	0
		Prisoners	0	3	0
		Imprisonment	0	0	0
		Imprisoned	0	0	0
		Jail	0	2	0
		Jailed	0	1	0
		Torture	0	3	1
		Tortures	0	0	2
		Tortured	0	5	0
		Torturing	0	1	1
		Riot	0	1	0
		Riots	1	2	0
		Fundamentalism	2	0	0
		Fundamentalist	0	2	1
		Fundamentalists	1	0	0
	Theme TV total		47	55	50
5.	Scientificity (S)				
		Science	19	0	0
		Sciences	1	0	0
		Scientist	0	0	0
		Scientists	5	0	0
		Scientific	6	0	0
		Technology	11	0	2
		Technologies	1	0	0
		Technological	3	0	0
		Technical	0	0	0
		Technically	0	0	0
		Technique	2	0	0
		Technologists	0	0	0
		Medicine	1	0	0
		Medical	3	0	0
		Astronomy	4	0	0
		Astronomer	0	0	0
		Physics	4	0	0

(continued on next page)

TABLE E1. (*continued*)

No.	Theme	Subtheme	Hindu Nationalism	Sikh Nationalism	Kashmiri Nationalism
		Physiology	1	0	0
		Physicist	0	0	0
		Physician	0	0	0
		Physicians	1	0	0
		Chemistry	3	0	0
		Chemical	3	0	1
		Chemist	0	0	0
		Engineering	9	0	0
		Engineers	1	0	0
		Engineer	0	0	0
		Engineered	0	0	0
		Metallurgy	0	0	0
		Nuclear	0	0	12
		Mathematics	5	0	0
		Mathematical	0	0	0
		Maths	0	0	0
	Theme S total		83	0	15
6.	Universality (U)				
		Universal	7	0	0
		Universally	0	0	3
		Universality	2	0	0
	Theme U total		9	0	3
7.	Truth and Knowledge (TK)				
		Truth	31	0	0
		Truths	4	0	0
		Knowledge	17	0	0
		Wisdom/ Wisdom-laden	21	0	3
	Theme TK total		73	0	3

Note. Sample size for nationalisms: HN N = 36, SN N = 15, KN N = 21.

Bibliography

"About HSC." *Hindu Students Council*, n.d. http://www.hscnet.org/aboutus.php (accessed January 4, 2006).

"About Us." *Hindu Unity*, n.d. http://hinduunity.org/aboutus.html (accessed March 19, 2006).

Adas, Michael. "Scientific Standards and Colonial Education in British India and French Senegal." In *Science, Medicine and Cultural Imperialism*, edited by Teresa Meade and Mark Walker, 4–35. New York: St. Martin's, 1991.

Aggarwal, J. C. *Documents on Higher Education in India (1781 to 1992): Selected Educational Statistics: World Overview*. Delhi: Doaba House, 1993.

Agnes, Flavia. *Law and Gender Inequality: The Politics of Women's Rights in India*. New Delhi: Oxford University Press, 1999.

Archer, Margaret. "Process Without System." In *Pierre Bourdieu*, edited by Derek Robbins, 1: 118–143. London: Sage, 2000.

Baber, Zaheer. *The Science of Empire: Scientific Knowledge, Civilization, and Colonial Rule in India*. Albany: State University of New York Press, 1996.

Balachandran, G., ed. *India and the World Economy*. New Delhi: Oxford University Press, 2003.

Bando, Ramen. *Secularism Extra-Ordinary*. London: Sahitya Porishad, 1997. http://www.hvk.org/specialrepo/seo/index.html (accessed March 19, 2006).

Bandopadhyay, Sekhar. *From Plassey to Partition: A History of Modern India*. Delhi: Orient Longman, 2004.

Bardhan, Pranab K. *The Political Economy of Development in India*. New York: Basil Blackwell, 1984.

Bhargava, Rajeev. "The Cultural Nationalism of the New Hindu." *Dissent* (Fall 2003): 11–17.

———, ed. *Secularism and Its Critics*. New Delhi: Oxford University Press, 1998.

Bhatt, Chetan. *Hindu Nationalism: Origins, Ideologies and Modern Myths*. New York: Berg, 2001.

Bidwai, Praful. "Plebiscite on Prejudice?" *Frontline* 16, no. 19, September 11–19, 1999: http://www.hinduonnet.com/fline/fl1619/16191030.htm (accessed March 19, 2006).

Birodkar, Sudheer. *Ancient India's Contribution to Our World's Material (Temporal) Culture*, n.d. http://www.geocities.com/Lavlesh/sudheer_contributions/index.html (accessed March 19, 2006).

Bose, Sumantra. " 'Hindu Nationalism' and the Crisis of the Indian State: A Theoretical Perspective." In *Nationalism, Democracy and Development: State and Politics*, edited by Sugata Bose and Ayesha Jalal, 104–164. New Delhi: Oxford University Press, 1997.

Bourdieu. Pierre. "Cultural Reproduction and Social Reproduction." In *Power and Ideology in Education*, edited by J. Karabel and A. H. Halsey, 487–511. New York: Oxford University Press, 1977.

———. *Distinction: A Social Critique of the Judgement of Taste*. Translated by Richard Nice. Cambridge, MA: Harvard University Press, 1984.

———. "Genesis and Structure of the Religious Field." *Comparative Social Research* 13 (1991): 1–44.

———. *Homo Academicus*. Translated by Peter Collier. Stanford, CA: Stanford University Press, 1988.

———. *Language and Symbolic Power*. Edited and introduced by John B. Thompson. Translated by Gino Raymond and Matthew Adamson. Cambridge, MA: Harvard University Press, 1991.

———. "The Myth of 'Globalization' and the Welfare State." In *Acts of Resistance: Against the Tyranny of the Market*, 29–45. New York: The New Press, 1998.

———. "Neo-liberalism, the Utopia (Becoming a Reality) of Unlimited Exploitation." In *Acts of Resistance: Against the Tyranny of the Market*, 94–105. New York: The New Press, 1998.

———. *Outline of a Theory of Practice*. Translated by Richard Nice. Cambridge, U.K.: Cambridge University Press, 1977.

———. *Pascalian Meditations*. Translated by Richard Nice. Stanford, CA: Stanford University Press, 1997.

———. *Practical Reason: On the Theory of Action*. Translated by Richard Nice. Stanford, CA: Stanford University Press, 1998.

———. *The Rules of Art: Genesis and Structure of the Literary Field*. Translated by Susan Emanuel. Stanford, CA: Stanford University Press, 1996.

———. *Science and Science of Reflexivity*. Translated by Richard Nice. Cambridge, U.K.: Polity, 2004.

———. *The State Nobility: Elite Schools in the Field of Power*. Translated by Lauretta C. Clough. Stanford, CA: Stanford University Press, 1996.

———. "What Makes a Social Class? On the Theoretical and Practical Existence of Groups." *Berkeley Journal of Sociology* 32 (1987): 1–17. Quoted in Ben Fine, *Social Capital Versus Social Theory: Political Economy and Social Science at the Turn of the Millennium*, 56. New York: Routledge, 2001.

Bourdieu, Pierre, and Luc Boltanski. "Changes in Social Structure and Changes in the Demand for Education." In *Contemporary Europe: Structural Change and Cultural Patterns*, edited by S. Giner and M. S. Archer, 197–227. London: Routledge & Kegan Paul, 1978.

Bourdieu, Pierre, and Jean-Claude Passeron. *Reproduction in Education, Society and Culture*. London: Sage, 1977.

Brass, Paul R. *The Politics of India since Independence*, 2nd ed. *The New Cambridge History of India*, part 4, vol. 1. Cambridge, U.K.: Cambridge University Press, 1994.

Brown, Judith M. *Nehru: A Political Life*. New Haven: Yale University Press, 2003.

Brubaker, Rogers. "Rethinking Classical Theory: The Sociological Vision of Pierre Bourdieu." In *Pierre Bourdieu*, edited by Derek Robbins, 3: 87–116. London: Sage, 2000.

Bunsha, Dionne. "The Facts from Godhra." *Frontline* 19, no. 15, July 20–August 2, 2002. http://www.frontlineonnet.com/fl1915/19150110.htm (accessed July 10, 2003).

Cafepress.com, n.d. http://www.cafepress.com/cp/info/ (accessed February 11, 2005).

Calhoun, Craig. "Introduction." *Comparative Social Research* 13 (1991): ix–xvi.

———. "Putting the Sociologist in the Sociology of Culture: The Self-Reflexive Scholarship of Pierre Bourdieu and Raymond Williams." In *Pierre Bourdieu*, ed. Derek Robbins, 3: 59–65. London: Sage, 2000.

Campaign to Stop Funding Hate. "The Cover-Up Begins: National HSC Goes One Step Further to Dupe Its Membership," *Campaign to Stop Funding Hate*, n.d. http://hsctruthout.stopfundinghate.org/PRs/pr_Question2.html (accessed September 4, 2008).

———. "Unmistakably Sangh: The National HSC and Its Hindutva Agenda." *Campaign to Stop Funding Hate*, n.d. http://hsctruthout.stopfundinghate.org/ (accessed September 4, 2008).

Castells, Manuel. *The Internet Galaxy: Reflections on the Internet, Business and Society*. Oxford: Oxford University Press, 2001.

———. *The Rise of the Network Society*, 2nd ed. Oxford: Blackwell, 2000.

Cecchini, Simone, and Monica Raina. "Warana: The Case of an Indian Rural Community Adopting ICT," April 2002, 1–9. Information

Technology in Developing Countries Working Paper. http://ssrn.com/abstract=568181 (accessed October 13, 2008).

Cerny, Philip G., Georg Menz, and Susanne Soederberg. "Different Roads to Globalization: Neoliberalism, the Competition State, and Politics in a More Open World." In *Internalizing Globalization: The Rise of Neoliberalism and the Decline of National Varieties of Capitalism*, edited by Susanne Soederberg, Georg Menz, and Philip G. Cerny, 1–30. New York: Palgrave Macmillan, 2005.

Chakrabarty, Dipesh. *Habitations of Modernity: Essays in the Wake of Subaltern Studies*. Chicago: University of Chicago Press, 2003.

Chakravartty, Paula. "The Emigration of High-Skilled Indian Workers to the United States: Flexible Citizenship and India's Information Economy." In *The International Migration of the Highly Skilled: Demand Supply, and Development Consequences in Sending and Receiving Countries*, edited by Wayne A. Cornelius, Thomas J. Espenshade, and Idean Salehyan, 325–349. La Jolla, CA: Center for Comparative Immigration Studies, 2001.

Chatterjee, Partha. "Development Planning and the Indian State." In *The State, Development Planning and Liberalisation in India*, edited by Terence J. Byres, 82–103. New Delhi: Oxford University Press, 1997.

———. *The Nation and Its Fragments: Colonial and Postcolonial Histories*. Princeton: Princeton University Press, 1993.

Chopra, Rohit. "Global Primordialities: Virtual Identity Politics in Online Hindutva and Online Dalit Discourse." *New Media and Society* 8, no. 2 (2006): 187–206.

———. "The Virtual State of the Nation: Online Hindu Nationalism in Global Capitalist Modernity." In *South Asian Technospaces*, edited by Radhika Gajjala and Venkataramana Gajjala, 153–178. New York: Peter Lang, 2008.

Clausen, Sten-Erik. *Applied Correspondence Analysis: An Introduction*. Thousand Oaks, CA: Sage, 1998.

Communalism Combat, n.d. http://www.sabrang.com/ (accessed August 10, 2005).

"ConcApp Concordancing Programs." *Edict Virtual Language Center*, n.d. http://www.edict.com.hk/pub/concapp/ (accessed April 3, 2005).

Corbridge, Stuart, and John Harriss. *Reinventing India: Liberalization, Hindu Nationalism and Popular Democracy*. Malden, MA: Blackwell, 2000.

Das, Arvind N. "The End of Geography: Nationalism in the Era of Globalization." In *Nations Under Siege: Globalization and Nationalism in Asia*, edited by Roy Starr, 31–62. New York: Palgrave, 2002.

Davies, Scott. "Leaps of Faith: Shifting Currents in Critical Sociology of Education." *American Journal of Sociology* 100, no. 6 (May 1995): 1448–1478.

Davis, Mike. *Late Victorian Holocausts: El Niño Famines and the Making of the Third World*. New York: Verso, 2001.

De Graaf, Peter. "The Impact of Financial and Cultural Resources on Educational Attainment in the Netherlands." *Sociology of Education* 59, no. 4 (October 1986): 237–246.

Deleuze, Gilles, and Félix Guattari. *Nomadology: The War Machine*. Translated by Brian Massumi. New York: Semiotext(e), 1986.

Deol, Harnik. *Region and Nationalism in India: The Case of the Punjab*. New York: Routledge, 2000.

Deol, Rajesh. "Cry for Khalistan Again." *Deccan Herald*, January 18, 2004. http://www.deccanherald.com/deccanherald/jan182004/fp5.asp (accessed May 4, 2005).

Diwanji, Amberish K. "Why Do These Hindu Groups Always Seek to Turn India Into a Hindu Pakistan?" *rediff.com*, February 4, 1999. http://www.rediff.com/news/1999/feb/04akd.htm (accessed March 11, 2005).

Doring, Nicola. "Personal Home Pages on the Web: A Review of Research." *Journal of Computer-Mediated Communication* 7, no. 3

(April 2002). http://jcmc.indiana.edu/vol7/issue3/doering.html (accessed July 7, 2005).

Du Gay, Paul, Stuart Hall, Linda Janes, Hugh Mackay, and Keith Negus. *Doing Cultural Studies: The Story of the Sony Walkman*. London: The Open University, 1997.

Dutt, Ela. "28.4 Million Foreign-Born Residents in U.S." *Indian-American Center for Political Awareness*, n.d. http://www.iacfpa.org/p_news/nit/iacpa-archieve/misc/24052002-census-1.html (accessed February 17, 2005).

Ebo, Bosah. "Internet or Outernet?" In *Cyberghetto or Cybertopia? Race, Class, and Gender on the Internet*, edited by Bosah Ebo, 1–12. Westport: Praeger, 1998.

Elst, Koenraad. *The Koenraad Elst Site*. http://koenraadelst.voiceofdharma.com/ (accessed June 17, 2005).

———. "The Politics of the Aryan Invasion Debate." *The Koenraad Elst Site* (April 2003). http://koenraadelst.bharatvani.org/articles/aid/aryanpolitics.html (accessed March 18, 2006).

Embassy of India, Washington DC. "Simla Agreement." *Embassy of India*. http://www.indianembassy.org/policy/Kashmir/shimla.htm (accessed June 10, 2005).

Engler, Steven. "Modern Times: Religion, Consecration and the State in Bourdieu." *Cultural Studies* 17, no. 3/4 (May/July 2003): 445–467.

Erikson, Bonnie. "Culture, Class, and Connections." *American Journal of Sociology* 102, no. 1 (July 1996): 217–251.

Fine, Ben. *Social Capital Versus Social Theory: Political Economy and Social Science at the Turn of the Millennium*. New York: Routledge, 2001.

Frankel, Francine. "Middle Classes and Castes in India's Politics: Prospects for Political Accommodation." In *India's Democracy: An Analysis of Changing State-Society Relations*, edited by Atul Kohli, 225–261. Princeton: Princeton University Press, 1988.

Friedman, Thomas. "Globalization, Alive and Well." *New York Times*, September 22, 2002, sec. 4, 13.

Frow, John. "Accounting for Tastes: Some Problems in Bourdieu's Sociology of Culture." In *Pierre Bourdieu*, edited by Derek Robbins, 3: 44–58. London: Sage, 2000.

Gandhi, M. K. *Hind Swaraj and Other Writings*. Edited by Anthony J. Parel. Cambridge, U.K.: Cambridge University Press, 1999.

Ganguly, Sumit. "Explaining the Kashmir Insurgency: Political Mobilization and Institutional Decay." *International Security* 21, no. 2 (Autumn 1996): 76–107.

Ganpati, Priya, and Ranjan Basu. "Pravasi Bharatiya Divas: Economy and the Diaspora." *rediff.com*, January 12, 2004. http://specials.rediff.com/money/2004/jan/10sld1.htm (accessed February 9, 2005).

Garnham, Nicholas, and Raymond Williams. "Pierre Bourdieu and the Sociology of Culture: An Introduction." In *Pierre Bourdieu*, edited by Derek Robbins, 3: 200–218. London: Sage, 2000.

Gartman, David. "Culture as Class Symbolization or Mass Reification? A Critique of Bourdieu's *Distinction*." *The American Journal of Sociology* 97, no. 2 (September 1991): 421–447.

Global Hindu Electronic Networks. "About This Site." *Hindu Universe: Hindu Resource Center*, n.d. http://www.hindunet.com/contact.htm (accessed February 19, 2005).

———. "Conversions." *Hindu Universe: Hindu Resource Center*, n.d. http://conversions.hindunet.org/ (accessed February 12, 2005).

———. "Dharma & Philosophy." *Hindu Universe: Hindu Resource Center*, n.d. http://www.hindunet.org/dharma/ (accessed February 11, 2005).

———. "History." *Hindu Universe: Hindu Resource Center*, n.d. http://www.hindunet.org/hindu_history/hindu_history.html (accessed February 12, 2005).

———. "In the Islamic Ages." *Hindu Universe: Hindu Resource Center*, n.d. http://www.hindunet.org/hindu_history/modern/moghal_link.html (accessed February 12, 2005).

———. "Interfaith Relations." *Hindu Universe: Hindu Resource Center*, n.d. http://www.hindunet.org/interfaith_relations/ (accessed February 12, 2005).

———. "Relationship Between Hindu and Moslem." *Hindu Universe: Hindu Resource Center*. http://www.hindunet.org/hindu_moslem/ (accessed February 12, 2005).

———. "Sciences." *Hindu Universe: Hindu Resource Center*, n.d. http://www.hindunet.org/sciences (accessed September 17, 2008).

Golwalkar, M. S. *We or Our Nationhood Defined*, 3rd ed. Nagpur: Bharat Prakashan, 1945.

Google, n.d. http://www.google.com (accessed May 1, 2005).

"Google Corporate Information, Technology Overview." *Google*, n.d. http://www.google.com/corporate/tech.html (accessed April 4, 2005).

"Google Information for Webmasters." *Google*, n.d. http://www.google.com/webmasters/4.html (accessed April 4, 2005).

Government of India. Ministry of External Affairs, Non Resident Indians and Persons of Indian Origin Division. "Estimated Size of Overseas Indian Community: Country-Wise." *Report of the High Level Committee on the Indian Diaspora*, December 2001. http://indiandiaspora.nic.in/diasporapdf/part1-est.pdf (accessed February 11, 2005).

———. "Foreword." *Report of the High Level Committee on the Indian Diaspora*, December 2001. http://indiandiaspora.nic.in/diasporapdf/part1-for.pdf (accessed February 11, 2005).

———. *The Indian Diaspora*, last updated October 5, 2004. http://indiandiaspora.nic.in (accessed February 11, 2005).

———. "Mandate of the Committee." *The Indian Diaspora*, n.d. http://indiandiaspora.nic.in/mandate.htm (accessed February 11, 2005).

Government of India. Ministry of Labour. "Industrial Development & Progress After Independence." In *Main Report*. Vol. 1, chap. 3, *Report of the Second National Commission on Labour*, 2002. http://www.labour.nic.in/lcomm2/2nlc-pdfs/Chap3.pdf (accessed May 11, 2004).

Government of India. Office of the Registrar General and Census Commissioner, India. "Census and You—Religion." *Census of India*, n.d. http://www.censusindia.gov.in/Census_And_You/religion.aspx (accessed June 22, 2008).

———. "Census Data 2001—National Summary Data Page (NSDP)." *Census of India*, n.d. http://www.censusindia.gov.in/Census_Data_2001/National_Summary/National_Summary_DataPage.aspx (accessed June 22, 2008).

———. "Census News—Data Highlights—New Releases." *eCENSUSIndia*, no. 4, November 30, 2001. http://www.censusindia.gov.in/Census_Data_2001/Census_Newsletters/Newsletter_Links/eci_4.htm (accessed June 22, 2008).

———. "Urban Population," *Census of India*, n.d. http://www.censusindia.gov.in/Census_Data_2001/Census_data_finder/A_Series/Urban_population.htm (accessed June 22, 2008).

Greenacre, Michael. "Correspondence Analysis and Its Interpretation." In *Correspondence Analysis in the Social Sciences: Recent Developments and Applications*, edited by Michael Greenacre and Jorg Blasius, 3–22. New York: Harcourt Brace, 1994.

———. *Correspondence Analysis in Practice*. San Diego: Academic Press, 1993.

Greenspan, Anna. *India and the IT Revolution: Networks of Global Culture*. New York: Palgrave, 2004.

Grover, B. L. and S. Grover. *A New Look at Modern Indian History: From 1707 to the Modern Times*. Delhi: S. Chand, 1988.

Grusky, David, and Jesper B. Sorensen. "Can Class Analysis Be Salvaged?" *The American Journal of Sociology* 103, no. 5 (March 1998): 1187–1234.

Guha, Ranajit. *Dominance without Hegemony: History and Power in Colonial India*. Cambridge, MA: Harvard University Press, 1997.

———. *An Indian Historiography of India: A Nineteenth-Century Agenda and Its Implications*. Calcutta, India: Center for Studies in Social Sciences, 1988.

———. "On Some Aspects of the Historiography of Colonial India." In *Selected Subaltern Studies*, edited by Ranajit Guha and Gayatri Chakravorty Spivak, 37–44. New York: Oxford University Press, 1988.

Guha, Ray Shantanu, Neerja Pawha Jetley, Charubala Annuncio, A.S. Panneerselvan, Nitin A. Gokhale, and Vasantha Arora. "The Best and the Brightest." *Outlook India*, May 29, 2000. http://www.outlookindia.com/full.asp?fname=cover_story&fodname=20000529&sid=1 (accessed February 11, 2004).

Habib, Irfan. "Introduction." In *The Eighteenth Century in Indian History. Evolution or Revolution?* edited by. P. J. Marshall, 1–49. New Delhi: Oxford University Press, 2003.

———. "Studying a Colonial Economy—Without Perceiving Colonialism." *Modern Asian Studies* 19, no. 3 (1985): 355–385.

Hansen, Thomas Blom. *The Saffron Wave: Democracy and Hindu Nationalism in Modern India*. Princeton: Princeton University Press, 1999.

Hardgrave, Robert, and Stanley Kochanek. *India: Government and Politics in a Developing Nation*. Fort Worth: Harcourt Brace, 1993.

Harris-White, Barbara. *India Working: Essays on Society and Economy*. Cambridge, U.K.: Cambridge University Press, 2003.

Himalayan Academy. "In Science and Beyond—*Vijnane Paratashcha*." In *Loving Ganesha: Hinduism's Endearing Elephant-Faced God*. 2nd ed. United States: Himalayan Academy, 2000. http://www.himalayanacademy.com/resources/books/lg/lg_ch-04.html (accessed March 19, 2006).

Hindman, Matthew, Kostas Tsioutsiouliklis, and Judy A. Johnson. *"Googlearchy": How a Few Heavily-Linked Sites Dominate the*

Web (2003). http://www.cs.princeton.edu/~kt/mpsa03.pdf (accessed October 12, 2007).

Hindu Holocaust Museum, n.d. http://www.mantra.com/holocaust/ (accessed July 2, 2005).

"Hindu Unity's Criminals Hit List: The Black List, Enemies of Hinduism Exposed," *Hindu Unity*. http://hinduunity.org/hitlist.html (accessed March 21, 2006).

Hindu Vivek Kendra, n.d. http://www.hvk.org/ (accessed March 19, 2006).

Hindugifts, n.d. http://hindunet.tolshop.com/ (accessed February 11, 2005).

Hindunet Signature Merchandise, n.d. http://www.cafepress.com/hindunet2 (accessed February 11, 2005).

Hindutva Brotherhood: Kashmiri Hindu Liberation Army. http://www.geocities.com/jairama1/and http://www.hindurashtra.org/ (accessed March 17, 2006).

"History Writing in India and Dangers From the Far Right." *South Asia Citizens Web*, n.d. http://sacw.insaf.net/India_History/index.html (accessed July 10, 2004).

Hofstede, Geert. *Culture's Consequences: International Differences in Work-Related* Values. Thousand Oaks, CA: Sage, 1980.

Holt, Richard. *Dialogue on the Internet: Language, Civic Identity and Computer-Mediated Communication*. Westport, CT: Praeger, 2004.

Honneth, Axel. "The Fragmented World of Social Forms: Reflections on Pierre Bourdieu's Sociology of Culture." In *Pierre Bourdieu*, edited by Derek Robbins, 3: 3–19. London: Sage, 2000.

Howard, Philip N. "Embedded Media: Who We Know, What We Know, and Society Online." In *Society Online: The Internet in Context*, edited by Philip. N. Howard and Steve Jones, 1–27. Thousand Oaks: Sage Publications, 2004.

"HSC Chapters." *Hindu Students Council*, n.d. http://www.hscnet.org/chapters.php (accessed January 4, 2006).

Huang, Philip C. C. "Rural Class Struggle in the Chinese Revolution: Representational and Objective Realities from the Land Reform to the Cultural Revolution." *Modern China* 21, no. 1 (January 1995): 105–143.

Human Rights Watch. *Behind the Kashmir Conflict: Abuses by Indian Security Forces and Militant Groups Continue*, July 1999. http://www.hrw.org/reports/1999/kashmir/abuses.htm (accessed June 1, 2005).

———. *India: Arms and Abuses in Indian Punjab and Kashmir*, 6, no. 10 (September 1994). http://www.hrw.org/campaigns/kashmir/1994/ (accessed June 11, 2005).

———. *India's Secret Army in Kashmir: New Patterns of Abuse Emerge in the Conflict*, 8, no. 4 (C) (May 1996). http://www.hrw.org/campaigns/kashmir/1996/ (accessed June 10, 2005).

"India: Internet Usage Stats and Telecommunications Market Report." *Internet World Stats*, n.d. http://www.internetworldstats.com/asia/in.htm (accessed June 19, 2007).

Institute of International Education. "International Student Enrollment in U.S. Rose 6.4% in 2001/2002, Matching Last Year's Increase as Highest Growth since 1980—India is the Top Sending Country." *IIE Online*, November 18, 2002. http://www.iie.org/PrinterTemplate.cfm?Section=News_Announcements&template=/ContentManagement/ContentDisplay.cfm&ContentID=5470 (accessed June 24, 2004).

Islam/Muslims in Kashmir, n.d. http://www.geocities.com/WestHollywood/Park/6443/Kashmir/ (accessed June 10, 2005).

"Islamic Terrorism and Genocide of Kashmiri Pandits." *Virtual Homeland of Kashmiri Pandits*, n.d. http://www.kashmiri-pandit.org/sundry/genocide.html (accessed July 3, 2005).

Joppke, Christian. "The Cultural Dimensions of Class Formation and Class Struggle: On the Social Theory of Pierre Bourdieu." In *Pierre Bourdieu*, edited by Derek Robbins, 3: 20–43. London: Sage, 2000.

Keay, John. *The Great Arc: The Dramatic Tale of How India Was Mapped and Everest Was Named.* London: HarperCollins, 2000.

Keer, Dhananjay. *Veer Savarkar*, 3rd ed. Bombay: Popular Prakashan, 1988.

Keniston, Kenneth. "Introduction: The Four Digital Divides." In *IT Experience in India: Bridging the Digital Divide*, edited by Kenneth Keniston and Deepak Kumar, 11–36. New Delhi: Sage Publications, 2004.

Khandelwal, Madhulika S. "Indian Immigrants in Queens, New York City: Patterns of Spatial Concentration and Distribution, 1965–1990." In *Nation and Migration: The Politics of Space in the South Asian Diaspora,* edited by Peter Van der Veer, 178–196. Philadelphia: University of Pennsylvania Press, 1995.

Khilnani, Sunil. *The Idea of India.* New Delhi: Penguin Books, 1999.

Kripalani, Manjeet, and Mark. K. Clifford. "India Wired." *Business Week*, Asian edition, February 21, 2000. http://www.businessweek.com/2000/00_08/b3669008.htm (accessed June 17, 2003).

Krueger, Anne O., and Sajjid Chinoy. "The Indian Economy in Global Context." In *Economic Policy Reforms and the Indian Economy,* edited by Anne O. Krueger, 9–45. Chicago: University of Chicago Press, 2002.

Kumar, Sanjay. "India's Urban Middle Class." In *Values and Life Styles in Urban Asia: A Cross-Cultural Analysis and Sourcebook Based on the AsiaBarometer Survey of 2003*, edited by Takashi Inoguchi, Miguel Basanez, Akihiko Tanaka, and Timur Dadabaev, 163–182. Mexico City: SigloXXI Editores, 2005.

Lal, Deepak. *Cultural Stability and Economic Stagnation: India c1500 BC–AD 1980.* Oxford: Clarendon Press, 1988.

Lal, Vinay. *The History of History: Politics and Scholarship in Modern India.* New Delhi: Oxford University Press, 2003.

———. "The Politics of History on the Internet: Cyber-Diasporic Hinduism and the North American Hindu Diaspora." In *In Diaspora: Theories, Histories, Texts*, edited by M. R. Paranjpe, 179–221. New Delhi: Indialog Publications, 2001.

Lamont, Michèle, and Anette Lareau. "Cultural Capital: Allusions, Gaps and Glissandos in Recent Theoretical Developments." *Sociological Theory* 6, no. 2 (Autumn 1988): 153–168.

Le Hir, Marie-Pierre. "Doing Cultural Studies Bourdieu's Way." In *Pierre Bourdieu*, edited by Derek Robbins, 3: 257–272. London: Sage, 2000.

"Links." *World Sikh Council.* http://www.worldsikhcouncil.org/docs/links_sites.html. Internet (accessed April 7, 2005).

Londhe, Sushama. "Advanced Concepts of Hinduism." *A Tribute to Hinduism*, updated September 7, 2005. http://www.atributetohinduism.com/Advanced_Concepts.htm (accessed March 19, 2006).

Macaulay, Thomas Babington. "Minute on Indian Education." In *Archives of Empire Volume 1: From the East India Company to the Suez Canal*, edited by Mia Carter and Barbara Harlow, 227–238. Durham: Duke University Press, 2003.

Madan, T. N. "Religions of India: Plurality and Pluralism." In *Religious Pluralism in South Asia and Europe*, edited by Jamal Malik and Helmut Reifeld, 42–76. New Delhi: Oxford University Press, 2005.

Marschall, Daniel. "Ideological Discourses in the Making of Internet Career Sites." *Journal of Computer-Mediated Communication* 7, no. 4 (July 2002). http://jcmc.indiana.edu/vol7/issue4/marschall.html (accessed July 10, 2005).

Mazumdar, Sucheta. "The Politics of Religion and National Origin: Rediscovering Hindu Indian Identity in the United States." In *Antinomies of Modernity: Essays on Race, Orient, Nation*, edited by Vasant Kaiwar and Sucheta Mazumdar, 223–260. Durham: Duke University Press, 2003.

McClelland, Katherine. "Cumulative Disadvantage Among the Highly Ambitious." *Sociology of Education* 63, no. 2 (April 1990): 102–121.

McDonald, Ellen E. "English Education and Social Reform in Late Nineteenth Century Bombay: A Case Study in the Transmission of a Cultural Ideal." *The Journal of Asian Studies*, 25, no. 3 (1966): 453–470.

McDowell, Stephen. *Globalization, Liberalization and Policy Change: A Political Economy of India's Communications Sector*. New York: St Martin's, 1997.

McLane, John. "The Drain of Wealth and Indian Nationalism at the Turn of the Century." In *India and the World Economy 1850–1950*, edited by G. Balachandran, 70–92. New Delhi: Oxford University Press, 2003.

Miller, Daniel, and Don Slater. *The Internet: An Ethnographic Approach*. Oxford, U.K.: Berg, 2000.

Mobilehindu, n.d. http://www.mobilehindu.com (accessed February 11, 2005).

"The Modi Factor." *The Hindu*, June 27, 2004. http://www.hindu.com/2004/06/27/stories/2004062700731400.htm (accessed March 19, 2006).

Mody, Nawaz B. "The Press in India: The Shah Bano Judgment and Its Aftermath." *Asian Survey* 27, no. 8 (August 1987): 935–953.

Morris, Sebastian. "Overview." In *India Infrastructure Report 2002: Governance Issues for Commercialization*, edited by Sebastian Morris and Rajiv Shekhar, 1–15. New Delhi: 3i Network and Oxford University Press, 2002.

Morris, Sebastian, and Rajiv Shekhar, eds. *India Infrastructure Report 2002: Governance Issues for Commercialization*. New Delhi: 3i Network and Oxford University Press, 2002.

Muppidi, Himadeep. *The Politics of the Global*. Minneapolis: University of Minnesota Press, 2004.

Nair, Janaki. *Women and Law in Colonial India: A Social History*. New Delhi: Kali for Women, 1996.

Nanda, Meera. *Prophets Facing Backward: Postmodern Critiques of Science and Hindu Nationalism in India*. New Brunswick: Rutgers University Press, 2003.

Nandy, Ashis. "Introduction: Science as a Reason of State." In *Science, Hegemony and Violence: A Requiem for Modernity,* edited by Ashis Nandy, 1–23. Tokyo: United Nations University, 1988.

———. *The Savage Freud and Other Essays on Possible and Retrievable Selves*. In *Return from Exile*. New Delhi: Oxford University Press, 2004.

———. *Time Warps: Silent and Evasive Pasts in Indian Politics and Religion*. New Brunswick: Rutgers University Press, 2002.

———. *Traditions, Tyranny and Utopias: Essays in the Politics of Awareness*. New Delhi: Oxford University Press, 1987.

"National Self Determination." *Khalistan, the New Global Reality*. http://www.khalistan.net/washtimes.htm (accessed June 10, 2005).

Needham, Anuradha Dingwaney, and Rajeswari Sunder Rajan, eds. *The Crisis of Secularism in India*, Durham: Duke University Press, 2007.

Nehru, Jawaharlal, "From Address to Youth, All-Bengal Students Conference—Calcutta, September 22." In *Nehru: The First Sixty Years*, selected and edited by Dorothy Norman, 166. London: The Bodley Head, 1965.

Nehru, Jawaharlal. "An Address at the Indian Science Congress, Calcutta, 26 December 1937." Speech documented in *Jawaharlal Nehru: An Anthology*, edited by Sarvepalli Gopal, 442. New Delhi: Oxford University Press, 1983.

Nehru, Jawaharlal. *The Discovery of India*. Calcutta: Signet Press, 1946. Reprint, New Delhi: Jawaharlal Nehru Memorial Fund and Oxford University Press, 1995.

———. *Nehru: The First Sixty Years*. Selected and edited by Dorothy Norman. London: The Bodley Head, 1965.

Neufeldt, Ronald. "The Hindu Mahasabha and Gandhi." In *Indian Critiques of Gandhi*, edited by Harold Coward, 131–152. Albany: State University of New York Press, 2003.

"Nine Beliefs of Hinduism." *Himalayan Academy*, n.d. http://www.himalayanacademy.com/basics/nineb/index.html (accessed February 11, 2005).

Noorani, A. G., ed. *The Babri Masjid Question, 1528–2003: A Matter of National Honour*, vols. 1 and 2. New Delhi: Tulika Books, 2003.

"Number of Internet Subscribers in India (as on 13.03.1999 to 30.06.2005)." *IndiaStat.com*, n.d. http://www.indiastat.com/india/ShowData.asp?secid=395675&ptid=143&level=3 (accessed June 26, 2007).

"Number of Subscribers in Cellular (Including WLL) and Fixed Line in India (2000 to 2005)." *Indiastat.com*, n.d. http://www.indiastat.com/india/ShowData.asp?secid=368487&ptid=19279&level=4 (accessed January 10, 2006).

Oberoi, Harjot. "From Punjab to 'Khalistan': Territoriality and Metacommentary." *Pacific Affairs* 60, no. 1 (Spring 1987): 26–41.

Ong, Aihwa. *Flexible Citizenship: The Cultural Logics of Transnationality*. Durham: Duke University Press, 1999.

Online Hindu Holocaust Museum, n.d. http://sarvadharma.org/Museum/HinduHolocaustMuseum.htm (accessed July 2, 2005).

"Our Search: Google Technology—Benefits of Google?" *Google*, n.d. http://www.google.com/technology/whyuse.html (accessed April 4, 2005).

"Our Search: Google Technology—Why Use Google?" *Google*, n.d. http://www.google.com/technology/index.html (accessed April 4, 2005).

Pan, Bing et al. "In Google We Trust: Users' Decisions on Rank, Position, and Relevance." *Journal of Computer-Mediated Communication* 12, no. 3, article 3 (2007). http://jcmc.indiana.edu/vol12/issue3/pan.html (accessed October 12, 2007).

Parashar, Archana. *Women and Family Law Reform in India*. Delhi: Sage Publishers, 1992. Quoted in Janaki Nair, *Women and Law in Colonial India: A Social History*, 181. New Delhi: Kali for Women, 1996.

Parekh, Bhikhu. *Gandhi's Political Philosophy: A Critical Examination*. London: Macmillan Press, 1989.

Persell, Caroline Hodges, and Peter W. Cookson Jr. "Microcomputers and Elite Boarding Schools: Educational Innovation and Social Reproduction." *Sociology of Education* 60, no. 2 (April 1987): 123–134.

Pileggi, Mary S., and Cindy Patton, "Introduction: Bourdieu and Cultural Studies." *Cultural Studies* 17, nos. 3/4 (May/July 2003): 313–325.

Poster, Mark. "Cyberdemocracy: The Internet and the Public Sphere." In *Virtual Politics: Identity and Community in Cyberspace*, edited by David Homes, 212–28. London: Sage, 1997.

Prakash, Gyan. *Another Reason: Science and the Imagination of Modern India*. Princeton: Princeton University Press, 1999.

———. "The Urban Turn." In *SARAI Reader '02: The Cities of Everyday Life*, edited by Ravi S. Vasudevan, Jeebesh Bagchi, Ravi Sundaram, Monica Narula, Geert Lovink and Shuddhabrata Sengupta, 2–7. New Delhi: Sarai Media Collective, 2002.

Prashad, Vijay. "No Entry for Modi." *Frontline* 22, no. 7, March 12–25, 2005. http://www.hinduonnet.com/fline/fl2207/stories/20050408003313100.htm (accessed March 12, 2006).

Progressive South Asians, n.d. http://www.proxsa.org/ (accessed August 10, 2005).

"Protest US Supported Marxist Assault against Hindus." *Petition Online*, n.d. http://www.petitiononline.com/108india/petition.html (accessed March 19, 2006).

Pseudo-Secularism, n.d. http://pseudosecularism.blogspot.com/ (accessed March 19, 2006).

Puri, Bindu. "Indian Nationalism and the Hindu Argument: Savarkar and Gandhi." In *Mahatma Gandhi and His Contemporaries*, edited by Bindu Puri, 154–176. Shimla: Indian Institute of Advanced Study, 2001.

Rajagopal, Arvind. "Hindu Nationalism in the US: Changing Configurations of Political Practice." *Ethnic and Racial Studies* 23, no. 3 (May 2000): 467–496.

———. *Politics After Television: Religious Nationalism and the Reshaping of the Indian Public*. Cambridge, U.K.: Cambridge University Press, 2001.

Rane, M. A. "Narendra Modi Denied Entry to the United States," *PUCL Bulletin*, August 2005. http://www.pucl.org/Topics/Religion-communalism/2005/modi-usa.htm (accessed March 19, 2006).

Rao, M. "India on Way to Digital Millennium, Madanmohan Rao Reports From the Gartner Summit in Mumbai, Aug. 14, 2000." *Inomy.com*. http://www.inomy.com/midd2.asp?storyid=1200134&sectioncode=17&subsectioncode=52&page=Subsection%20PageINOMY%20EXCLUSIVESE-COMMERCE. Internet. Quoted in Peter Wolcott and Seymour Goodman, "Global Diffusion of the Internet I: Is the Elephant Learning to Dance?" *Communications of the Association for Information Systems* 11 (2003): 612. http://cais.isworld.org/articles/11-32/default.asp?View=Journal&x=34&y=14 (accessed May 17, 2005).

Ray, Rajat Kanta. *The Felt Community: Commonality and Mentality Before the Emergence of Indian Nationalism*. New Delhi: Oxford University Press, 2003.

Rheingold, Howard. *The Virtual Community: Homesteading on the Electronic Frontier*, rev. ed. Cambridge, MA: MIT Press, 2000.

Rigby, Brian. "Popular Culture as Barbaric Culture: The Sociology of Pierre Bourdieu." In *Pierre Bourdieu*, edited by Derek Robbins, 3: 273–302. London: Sage, 2000.

Ringer, Fritz. "The Intellectual Field, Intellectual History, and the Sociology of Knowledge." In *Pierre Bourdieu*, edited by Derek Robbins, 3: 66–86. London: Sage, 2000.

Robbins, Derek. "Introduction." In *Pierre Bourdieu,* edited by Derek Robbins, 1: 1–65. London: Sage, 2000.

———, ed. *Pierre Bourdieu.* 4 vols. London: Sage, 2000.

Robinson, Robert V., and Maurice A.Garnier. "Class Reproduction among Men and Women in France: Reproduction Theory on Its Home Ground." *American Journal of Sociology* 91, no. 2 (September 1985): 250–280.

Rothermund, Dietmar. *An Economic History of India: From Pre-Colonial Times to 1991*, 2nd ed. New York: Routledge, 1993.

Sachs, J. D., A. Varshney, and N. Bajpai. "Preface to the Paperback edition." In *India in the Era of Economic Reforms*, edited by J. D. Sachs, A. Varshney, and N. Bajpai, vii–xvii. New Delhi: Oxford University Press, 2000.

Sahasrabudhey, Sunil. "*Hind Swaraj* and the Science Question." In *Hind Swaraj: A Fresh Look,* edited by Nageshwar Prasad, 99–105. New Delhi: Gandhi Peace Foundation, 1985.

Sangwan, Satpal. *Science, Technology and Colonisation: An Indian Experience 1757–1857*. New Delhi: Anamika Prakashan, 1991.

SARAI. n.d. http://www.sarai.net/ (accessed August 10, 2005).

Sardar, Ziauddin. "alt.civilizations.faq: Cyberspace as the Darker Side of the West." In *Cyberfutures: Culture and Politics on the Information Superhighway*, edited by Ziauddin Sardar and Jerome R. Ravetz, 14–41. London: Pluto Press, 1996.

Sarkar, Tanika. "Heroic Women, Mother Goddesses: Family and Organisation in Hindutva Politics." In *Women and Right-Wing Movements: Indian Experiences*, edited by Tanika Sarkar and Urvashi Butalia, 181–215. London: Zed Books, 1995.

Savarkar, Vinayak Damodar. *Hindu-Pad-Padshahi or a Review of the Hindu Empire of Maharashtra*, 4th ed. New Delhi: Bharatiya Sahitya Sadan, 1971.

———. *Hindutva: Who Is a Hindu?* 5th ed. Bombay: Veer Savarkar Prakashan, 1969.

———. *Historical Statements—V. D. Savarkar*. Edited by S. S. Savarkar and G. M. Joshi. Bombay: Popular Prakashan, 1967.

Sen, Sudipta. *Empire of Free Trade: The East India Company and the Making of the Colonial Marketplace*. Philadelphia: University of Pennsylvania Press, 1998.

Shah, Ghanshyam. "Tenth Lok Sabha Elections: BJP's Victory in Gujarat." *Economic and Political Weekly*, December 21, 1991, 2921–2924.

Shurmer-Smith, Pamela. *India: Globalization and Change*. London: Arnold, 2000.

"Sikh Struggle for Independence." *Khalistan, the New Global Reality*. http://www.khalistan.net/unpo.htm (accessed June 10, 2005).

Singh, Nitish, and Daniel W. Baack. "Web Site Adaptation: A Cross-Cultural Comparison of U. S. and Mexican Web Sites." *Journal of Computer-Mediated Communication* 9, no. 4 (July 2004). http://jcmc.indiana.edu/vol9/issue4/singh_baack.html (accessed July 11, 2005).

Singhal, Arvind, and Everett M. Rogers. *India's Communication Revolution: From Bullock Carts to Cyber Marts*. New Delhi: Sage Publications, 2001.

Sonia Maino, Sonia Gandhi. n.d. http://www.soniamaino.com/ (accessed March 20, 2006).

South Asia Citizens Web. n.d. http://www.sacw.net/ (accessed August 10, 2005).

Sulkunnen, Pekka. "Society Made Visible: On the Cultural Sociology of Pierre Bourdieu." In *Pierre Bourdieu*, edited by Derek Robbins, 3: 219–232. London: Sage, 2000.

Swartz, David. "Pierre Bourdieu: The Cultural Transmission of Social Inequality." In *Pierre Bourdieu*, edited by Derek Robbins, 2: 207–217. London: Sage, 2000.

Tangri, Shanti S. "Intellectuals and Society in Nineteenth-Century India." *Comparative Studies in Society and History* 3, no. 4 (1961): 368–394.

Tateo, Luca. "The Italian Extreme Right On-Line Network: An Exploratory Study Using an Integrated Social Network Analysis and Content Analysis Approach." *Journal of Computer-Mediated Communication* 10, no. 2 (January 2005). http://jcmc.indiana.edu/vol10/issue2/tateo.html (accessed July 10, 2005).

Thompson, John B. "Editor's Introduction." In Pierre Bourdieu, *Language and Symbolic Power*, edited and introduced by John B. Thompson, translated by Gino Raymond and Matthew Adamson, 1–31. Cambridge, MA: Harvard University Press, 1991.

"Timeline: Ayodhya Crisis." *BBC News*, U.K. edition, last updated July 5, 2005. http://news.bbc.co.uk/1/hi/world/south_asia/1844930.stm (accessed August 15, 2005).

Tremblay, Reeta Chowdhury. "Nation, Identity and the Intervening Role of the State: A Study of the Secessionist Movement in Kashmir." *Pacific Affairs* 69, no. 4 (Winter 1996–1997): 471–497.

Upadhyaya, Prakash Chandra. "The Politics of Indian Secularism." *Modern Asian Studies* 26, no. 4 (October 1992): 815–853.

Vanaik, Achin. "Consumerism and New Classes in India." In *Thinking Social Science in India: Essays in Honour of Alice Thorner*, edited by Sujata Patel, Jasordhara Bagchi, and Krishna Raj, 227–234. New Delhi: Sage Publications, 2002.

———. *The Furies of Indian Communalism: Religion, Modernity, and Secularization*. London: Verso, 1997.

———. *The Painful Transition: Bourgeois Democracy in India*. London: Verso, 1990.

Varshney, Ashutosh. "Understanding Gujarat Violence." *Contemporary Conflicts*, March 26, 2004. http://conconflicts.ssrc.org/gujarat/varshney/ (accessed September 23, 2008).

Vartak, P. V. "The Scientific Dating of the Mahabharat War." *Hindu Universe*, n.d. http://www.hindunet.org/hindu_history/ancient/mahabharat/mahab_vartak.html (accessed January, 11 2005).

Vedantam, Shankar. "Wrath Over a Hindu God: U.S. Scholars' Writings Draw Threats From Faithful." *Washington Post*, April 10, 2004, A01. http://www.washingtonpost.com/ac2/wp-dyn?pagename=article&contentId=A334-2004Apr9&notFound=true (accessed August 11, 2005).

Verter, Bradford. "Spiritual Capital: Theorizing Religion with Bourdieu Against Bourdieu." *Sociological Theory* 21, no. 2 (June 2001): 150–173.

Visvanathan, Shiv. *A Carnival for Science: Essays on Science, Technology, and Development*. Delhi: Oxford University Press, 1997.

———. *Organizing for Science: The Making of an Industrial Laboratory*. New Delhi: Oxford University Press, 1985.

Wacquant, Loic J. D. "Towards a Reflexive Sociology: A Workshop with Pierre Bourdieu." *Sociological Theory* 7, no. 1 (Spring 1989): 26–63.

"Who Are the Kashmir Militants?" *BBC News*, February 19, 2003. http://news.bbc.co.uk/1/hi/world/south_asia/1719612.stm#map (accessed July 17, 2005).

Wolcott, Peter, and Seymour Goodman. "Global Diffusion of the Internet I: Is the Elephant Learning to Dance?" *Communications of the Association for Information Systems* 11, no. 32 (2003): 560–646. http://cais.isworld.org/articles/11-32/default.asp?View=Journal&x=34&y=14 (accessed May 17, 2005).

World Sikh Council, n.d. http://www.worldsikhcouncil.org/ (accessed April 7, 2005).

"WSC-AR Condemns Denial of Justice to Sikh Victims of November 1984 in India; Appeals to World Community and UN." *World*

Sikh Council (October 28, 2004). http://www.worldsikhcouncil.org/press/2004/2004_Nov_PressRelease.pdf (accessed April 7, 2005).

Zastoupil, Lynn, and Martin Moir. "Introduction." In *The Great Indian Education Debate: Documents Relating to the Orientalist-Anglicist Controversy 1781–1843*, edited by Lynn Zastoupil and Martin Moir, 1–72. Richmond: Curzon, 1999.

Index

Made in the USA
Monee, IL
12 November 2026

f3bfa470-349b-4c55-af75-8506cd03a95cR01